Water Treatment Operator Training Handbook

Water Treatment Operator Training Handbook

Third Edition

Nicholas G. Pizzi
William C. Lauer

Disclaimer
This book is provided for informational purposes only, with the understanding that the publisher, editors, and authors are not thereby engaged in rendering engineering or other professional services. The authors, editors, and publisher make no claim as to the accuracy of the book's contents, or their applicability to any particular circumstance. The editors, authors, and publisher accept no liability to any person for the information or advice provided in this book or for loss or damages incurred by any person as a result of reliance on its contents. The reader is urged to consult with an appropriate licensed professional before taking any action or making any interpretation that is within the realm of a licensed professional practice.

ISBN: 978-1-58321-861-7

Project Manager/Senior Technical Editor: Melissa Valentine
Cover Design: Melanie Yamamoto
Production: Janice Benight Design Studio

Library of Congress Cataloging-in-Publication Data
Pizzi, Nicholas G.
[Water treatment operator handbook]
Water treatment operator training handbook / by Nicholas G. Pizzi and William C. Lauer. -- Third edition.
pages cm
Previous edition published as: Water treatment operator handbook, 2005.
Includes index.
ISBN 978-1-58321-861-7
1. Water treatment plants--Handbooks, manuals, etc. 2. Water--Purification--Handbooks, manuals, etc. I. Lauer, Bill. II. Title.
TD434.P579 2013
628.1'62--dc23

2013009995

Contents

First Edition Preface

The people who operate water treatment plants make an important contribution to the public health of the community in which they work. A safe drinking water supply is essential to all representative groups within the sphere of influence of the treatment plant. Therefore, the water plant operator should be a trained professional who is capable of performing the tasks necessary to provide the safest possible water.

Trained professional: the phrase implies two principles. First, this is a person who conforms to technical and ethical standards of a discipline, perhaps on the level of a permanent career. Second, this person is trainable and has been trained to perform at a level that leads to professionalism. This book is written with those principles in mind.

An ethical paradigm is in action here. Performing the role of water plant operator places a premium on allegiance to the consumer, the regulator, and the employer. Often, operators believe the demands made on them by the three are in conflict. Many operator training programs are available to people in the field, but few of them offer training in ethics. Occasionally, we read about an operator who has made a questionable ethical choice. Public response to these incidents is a predictable loss of confidence in the drinking water industry. Usually, when operators make the right choice, it is a result of their training. Unfortunately, the right choices that operators make are not often written about.

The US Environmental Protection Agency (USEPA) has published guidelines for the certification and qualification of the men and women who operate the water treatment plants in this country.* These guidelines provide a legal framework for operator certification, one that underscores the opinions previously stated. To add weight to the seriousness of the rule, the USEPA administrator is directed to withhold 20 percent of the funds a state is otherwise entitled to receive in its Drinking Water State Revolving Fund capitalization grants under section 1452 of the Safe Drinking Water Act if the state does not meet goals or commitments. This action is an inducement to perform.

*Federal Register, USEPA, FRL-6230-8, Final Guidelines for the Certification and Recertification of the Operators of Community and Nontransient Noncommunity Public Water Systems, Feb. 5, 1999. p. 5916.

The rule requires, at a minimum, that all owners of public water supplies place the direct supervision of their water system under the responsible charge of an operator(s) holding a valid certification equal to or greater than the classification of the treatment facility. It further requires that all operating personnel making process-control decisions about water quality or quantity that affect public health be certified. Finally, it requires that a designated certified operator be available for each operating shift.

The rule is populated with many phrases that begin with "must": the state must classify water systems; operators must pass an exam; exam questions must be validated; operators must have a minimum level of schooling; licenses must be renewed; the state must establish training requirements; the state must include ongoing stakeholder involvement.

What drives this rule and its associated requirements? The recognition that the water treatment plant is a barrier against the passage of pathogen-infested or otherwise harmful water and that those who own and operate them should be accountable for the water they produce. The concept of the multiple-barrier approach to water plant operation has evolved from such recognition. Programs such as the Partnership for Safe Water and other efforts attempt to instill a systematic approach to this discipline and to encourage tenacity for ongoing process improvement through optimization. These programs provide a framework for operators who wish to improve their skills and knowledge of the profession.

It is hoped that this third edition will help operators in their search for professionalism.

I would like to thank the following water utility professionals who served as reviewers of this book. Their time is appreciated and their expertise acknowledged.

Christine A. Owen, PhD, Water Quality Assurance Officer, Tampa Bay Water, Clearwater, Fla.

Michael J. Pickel, PE, BS Civil Engineering, Environmental Programs Manager, Philadelphia Water Department, Philadelphia, Pa.

David A. Visintainer, MSCE, Director of Public Utilities and Water Commissioner, City of St. Louis Water Division, St. Louis, Mo.

Melinda L. Raimann, BSEd, CUSA, Assistant Commissioner, Cleveland Division of Water, Cleveland, Ohio

David J. Rexing, BA Chemistry, MBA, Water Quality Research and Development Manager, Southern Nevada Water Authority, Las Vegas, Nev.

Jan C. Routt, BS Microbiology, Director of Water Quality, Kentucky-American Water Company, Lexington, Ky.

I would also like to thank the following individuals for their assistance in developing the outline for this book. Their contribution was thorough and helped to make this work a comprehensive tool for operators: David Talley, Bill Lauer, Joe McDonald, Nelson Yarlott, and Gay Porter De Nileon.

Nicholas G. Pizzi

* * *

Third Edition Preface

Nick Pizzi is my friend and colleague. When AWWA asked me to consider authoring the third edition of this book, the first thing I did was to contact Nick. He explained that he had decided not to participate in the revision and that he was confident that I would provide professional and current information.

Even though I did not work with Nick on this new edition, his hand is evident throughout. Before making his decision not to do this revision, he had provided suggestions for updates and I have included them. Also, I did not tamper with Nick's style. His words reveal a passion for operations excellence that I can only hope to appreciate.

My approach was to update the many references and illustrations to make them current. I also reviewed the operator certification knowledge requirements included in the Associated Boards of Certification (ABC) Need-to-Know criteria. Several state certification boards' (e.g., California, Pennsylvania, and Texas) operator requirements were also added where there were differences. Any topics in these certification requirements that were not already included in the book were added.

The result of these additions and revisions make this book an indispensable reference for all water treatment plant operators and plant managers. The book is equally useful for certification exam studies and as a reference for operations personnel.

I want to thank David Plank, Melissa Valentine, and Alan Roberson at AWWA for their help and support during the publication process. Most of all I want to thank Nick Pizzi for his outstanding book, which I was allowed to supplement.

William C. Lauer

Chapter 1

Regulated Contaminants and Treatment Challenges

Regulations that govern US water supply and treatment are developed by the US Environmental Protection Agency (USEPA) under the Safe Drinking Water Act (SDWA). Most states administer USEPA regulations after adopting regulations that are no less stringent than federal rules; and in some cases, states have adopted stricter regulations or have developed regulations for additional contaminants not regulated by USEPA.

This chapter discusses current and anticipated USEPA regulations and the challenges that operators face in their efforts to comply with the regulations. Water system operators should consult their local and state regulatory agencies to verify applicable regulations that may be different than the federal regulations listed in this chapter. The chapter concludes with a discussion of selected contaminants that are commonly found in water, their significance, and the methods for their removal.

TYPES OF WATER SYSTEMS

The SDWA defines a public water system (PWS) as a supply of piped water for human consumption that has at least 15 service connections, or serves 25 or more persons 60 or more days each year. By that definition, private homes, groups of homes with a single water source but having fewer than 25 residents, and summer camps with their own water source that operate less than 60 days per year are not PWSs. They may, however, be subject to state or local regulations. Such systems may also be subject to state and local well construction and water quality requirements.

PWSs are classified into three categories based on the type of customers served:

- *Community PWS:* a system whose customers are full-time residents
- *Nontransient noncommunity PWS:* an entity having its own water supply, serving an average of at least 25 persons who do not live at the location but who use the water for more than 6 months per year

- *Transient noncommunity PWS*: an establishment having its own water system, where an average of at least 25 people per day visit and use the water occasionally or for only short periods of time

The rationale for these classifications is based on the differences in exposure to contaminants experienced by persons using the water. Most chemical contaminants are believed to potentially cause adverse health effects from long-term exposure. Short-term exposure to low-level chemical contamination may not carry the same risk as long-term exposure.

Therefore, the monitoring requirements for both community and noncommunity water systems apply to all contaminants that are considered a health threat. The transient and nontransient noncommunity systems must only monitor for nitrite and nitrate, as well as biological contamination (those that pose immediate threat from brief exposure). The remaining community systems, about 52,000 in the United States, have more stringent and frequent monitoring requirements.

Before examining the specific regulations that govern contaminants, the operator needs to know the difference between the two concepts used in the contaminant monitoring process: the maximum contaminant level goal (MCLG) and the maximum contaminant level (MCL).

- The MCLG is set for most substances at a level where there are no known, or anticipated, health effects. For those substances that are suspected carcinogens, the MCLG is set at zero.
- The MCL is set as close as feasible to the MCLG for substances regulated under the SDWA. The MCL is a level that is reasonably and economically achievable. This is the enforceable regulated level. Water systems that exceed an MCL must take steps to install treatment to reduce the contaminant concentration to below the MCL. Where USEPA has found it impractical to set an MCL, a treatment technique (TT) has been established instead of an MCL.

With these concepts in mind, the various regulations can be examined. This discussion is not meant to be all-inclusive. Because the regulatory process is an ever-evolving one, the reader is cautioned that some of the stated facts presented in this discussion may have changed since the writing of this chapter. For up-to-date information, it is best to contact the local office of the regulatory authority in the district or state where the utility operates.

Table 1-1 contains some of the more common regulated contaminants and their respective MCL or treatment technique (TT) descriptions. These are provided for illustration only and are not intended to be used for regulatory purposes (see the official USEPA regulatory information on the agency website).

Operations personnel are expected to know the regulatory limits for compounds encountered in their water supply. However, the number and variety of regulated substances make it unlikely that operators would know all of the regulatory limits. Operators must rely on current references for the most accurate information. These are available from the regulatory agency responsible for the location of the treatment plant.

Table 1-1 Selected USEPA drinking water standards

Contaminant	MCL or TT (mg/L)*
Total coliform	5 percent (monthly positives)
Turbidity	0.3 ntu monthly or 1 ntu†
Chlorite	1.0
HAA5	0.060
TTHM	0.080
Chloramines (as Cl_2)	4.0
Chlorine (as Cl_2)	4.0
Chlorine dioxide (as ClO_2)	0.8
Arsenic	0.010
Copper	TT, Action level = 1.3
Cyanide (as free cyanide)	0.2
Fluoride	4.0
Lead	TT; Action Level = 0.015
Mercury (inorganic)	0.002
Nitrate (measured as nitrogen)	10
Nitrite (measured as nitrogen)	1
Radium 226 and 228 (combined)	5 pCi/L
Uranium	10 µg/L

*The listed standards are numerical representations of the current USEPA drinking water standard and do not include the sample frequency or location and other important compliance information. For a complete definition of the standards consult USEPA Drinking Water Standards.

†Turbidity less than or equal to 0.3 nephelometric turbidity units (ntu) for the combined filter effluent for 95% of the monthly samples. At no time can turbidity be above 1 ntu.

DISINFECTION BY-PRODUCT AND MICROBIAL REGULATIONS

Drinking water treatment, including use of chemical disinfectants such as chlorine, ozone, and chlorine dioxide, has been an important step in protecting drinking water consumers from exposure to harmful microbial contaminants. However, these chemical disinfectants can also react with organic and inorganic substances in the water to produce by-products that may be harmful to drinking water consumers, particularly some susceptible segments of the population. Therefore, drinking water treatment using chemical disinfectants involves a delicate balancing act, i.e., adding enough disinfectant to control harmful microorganisms but not enough to produce unacceptably high levels of regulated disinfection by-products (DBPs).

The USEPA has enacted several regulations impacting microbial control and production of DBPs in groundwater and surface water supplies for small and large public drinking water systems. These rules are referred to collectively as the Microbial/Disinfection By-Products (M/DBP) Rules. Microbial protection for consumers of drinking water from public supplies is provided by provisions of current or pending rules listed below and discussed in more detail later in this chapter:

- Filter Backwash Recycling Rule (FBRR)
- Ground Water Rule (GWR)
- Interim Enhanced Surface Water Treatment Rule (IESWTR)
- Long-Term 1 Surface Water Treatment Rule (LT1ESWTR)
- Long-Term 2 Surface Water Treatment Rule (LT2ESWTR)
- Stage 1 Disinfectants and Disinfection By-products Rule (Stage 1 DBPR)
- Stage 2 Disinfectants and Disinfection By-products Rule (Stage 2 DBPR)
- Surface Water Treatment Rule (SWTR)
- Total Coliform Rule (TCR)

Provisions of the Disinfectants and Disinfection By-products Rule (DBPR) are intended to protect drinking water consumers against the unintended public health consequences associated with consumption of treated drinking water containing residual disinfectants and DBPs produced from degradation of these residual disinfectants or reaction of disinfectants with organic and inorganic DBP precursors.

More details regarding the DBPR, including the current Stage 1 and 2 DBPR, are described in this chapter. Also included in the DBPR description is a brief discussion of some currently unregulated DBPs that are being heavily researched and may be the subject of future regulation. In the following discussion, the DBPR will be discussed first, followed by the microbial protection rules (SWTR, GWR, and TCR/DSR).

DBPR

The Stage 1 and 2 DBPR requirements discussed in the following sections focus first on two specific contaminants (TTHM and HAA5), and then on other aspects of these regulations dealing with control or removal of DBP precursors ("enhanced coagulation"), bromate, chlorite, and residual disinfectants.

Stage 1 DBPR—HAA5 and TTHM Provisions

The Stage 1 DBPR was published in 1998 and established an MCL of 0.080 mg/L for TTHM (the sum of four trihalomethanes, which are chloroform, bromodichloromethane, dibromochloromethane, and bromoform) and 0.060 mg/L for HAA5 (the sum of five specific haloacetic acids, which are mono-, di-, and tri-chloroacetic acids plus mono- and dibromoacetic acids). Although the MCLs for TTHM and HAA5 were officially written as 0.080 mg/L and 0.060 mg/L, respectively, the limits are commonly referred to as "80/60," or 80 µg/L and 60 µg/L. Although the numerical value of each MCL is an important consideration, an understanding of the methodology used to calculate the compliance value in order to compare it to this MCL is a subtle and equally important consideration in understanding compliance with the DBPR.

For TTHM and HAA5, the compliance value is determined by monitoring the distribution system. Compliance monitoring locations need to be representative of the distribution system. Systems serving >10,000 persons that use surface water sources are required to monitor at least four locations per plant, meaning that distribution systems fed by more than one treatment plant must have at least four monitoring locations designated for each plant entry point.

The compliance monitoring location for systems with only one monitoring point must be representative of maximum residence time in the distribution system. A minimum of one out of every four compliance monitoring locations for systems with more locations must also be representative of maximum residence time. The other locations must be far enough away from the plant entry points to be representative of average residence time in the distribution system.

Unlike acute toxicity risks, for which the exposure could be a single glass of water, cancer risks like those believed to be linked to TTHM and HAA5 involve longer periods of exposure (daily glasses of water spanning decades). For chronic exposures such as these, exposure to an excessively high concentration of a given cancer-causing agent will not necessarily result in the consumer getting cancer from this source. Conversely, a consumer exposed to a lower concentration every day for a lifetime could be more likely to develop cancer. Therefore, regulation of DBPs to reduce cancer risks is *not* based on limiting exposure to a single incident (i.e., not a "single hit"), but rather is aimed at reducing the repeated exposure over time. In other words, DBP exposure needs to be evaluated on an average basis over time.

Under the Stage 1 DBPR, the compliance value for TTHM and HAA5 is determined by calculating a running annual average (RAA) during the previous 12 months for each DBP for all monitoring locations at each plant. Most systems are required to monitor quarterly (i.e., 4 times per year), although small groundwater systems (<10,000 persons) may be allowed to sample once a year. Typically, the RAA is based on 4 monitoring locations sampled quarterly, meaning RAA will be the average of 16 monitoring results each for HAA5 and TTHM.

Table 1-2 illustrates one facility's calculations of RAA for HAA5 that were used for Stage 1 compliance (this table also shows calculation of values for Stage 2 DBPR, which will be discussed later). It is important to reemphasize that compliance is based solely on the RAA, not on a single quarterly result at any one monitoring location. Consequently, it is *not* correct to refer to a single quarterly monitoring result above 60 µg/L for HAA5 or above 80 µg/L for TTHM as being above the MCL. Therefore, even though several individual monitoring values in Table 1-2 are greater than 60 µg/L, the facility is in compliance with the HAA5 MCL because the RAA is 45 µg/L for HAA5.

Utility personnel should be consistent and rigorous in their use of terminology when dealing with the general public or with state and local health officials, and should ensure that all people participating in these discussions are consistent in applying the MCL only to RAA values and do not make the common mistake of referring to a single quarterly monitoring value as being "above the MCL."

Stage 2 DBPR—HAA5 and TTHM Provisions

The Stage 2 DBPR, published in 2006, is now in effect. This rule tightened requirements for DBPs, but compliance is not achieved by modifying the numerical value of the MCLs or by requiring monitoring of new constituents. Instead, the rule makes compliance more challenging than under the Stage 1 DBPR by (1) changing the way the compliance value is calculated and (2) changing the compliance monitoring locations to sites representative of the greatest potential for THM and HAA formation. These changes were made to ensure uniform compliance with the DBP standards across all areas of the distribution system, i.e., compliance is required at each sampling location.

The compliance value in the Stage 2 DBPR is called the *locational running annual average* (LRAA), and it is calculated by separately averaging the four quarterly samples at each monitoring location. Compliance is based on the maximum LRAA value (see Table 1-2). Furthermore, the Stage 2 DBPR included several interim steps that led to the replacement of many existing Stage 1 DBPR monitoring locations with new locations representative of the greatest potential for consumer exposure to high levels of TTHM and HAA5.

The Stage 2 DBPR required that facilities maintain compliance with the Stage 1 DBPR using the existing monitoring locations during the first three years after the final version of the Stage 2 DBPR was published. In the time period between the third and sixth year after the Stage 2 DBPR was published, compliance continues to be based on maintaining 80/60 (TTHM and HAA5) or lower for RAA; it also includes a requirement for maximum LRAA at existing Stage 1 monitoring locations. The long-term goal of the Stage 2 DBPR was to identify locations within the distribution system with the greatest potential for either TTHM or HAA5 formations and then base compliance on the LRAA at or below 80/60 for each of these locations. Many of these locations were identified during the initial distribution system evaluation (IDSE).

Table 1-2 Example RAA and LRAA calculations for Stage 1 and 2 DBPR

		Sampling Location, μg/L			
Year	**Quarter**	**A**	**B**	**C**	**D**
1	3rd	52	68	63	66
1	4th	35	42	38	41
2	1st	47	49	42	43
2	2nd	18	42	45	37
LRAA		38	50	47	47
Maximum LRAA			50		
RAA			45		

The IDSE included monitoring, modeling, and/or other evaluations of drinking water distribution systems to identify locations representative of the greatest potential for consumer exposure to high levels of TTHM and HAA5. The goal of the IDSE was to evaluate a number of potential monitoring locations to justify selection of monitoring locations for long-term compliance (i.e., Stage 2B) with the Stage 2 DBPR.

One item to note regarding the Stage 2 DBPR as it applies to TTHM and HAA5 is that the goal was to find the locations in the distribution system where average annual levels of these DBPs are highest. TTHM formation increases as contact time with free or combined chlorine increases, although formation in presence of combined chlorine is limited. Therefore, establishing points in the distribution system with highest potential for TTHM formation is related to points with maximum water age. Utilities that have not performed a tracer study in the distribution system to determine water age should consider doing so.

By contrast, peak locations for HAA5 are more complicated because microorganisms in biofilm attached to distribution system pipe surfaces can biodegrade HAA5. Consequently, increasing formation of HAA5 over time is offset by biodegradation, eventually reaching a point where HAA5 levels decrease over time, even to the point where they drop to zero. Figure 1-1 shows gradual reduction in HAA5 formation over time in a distribution system, followed by eventual decrease of HAA5 as water age increases (water age measured in tracer test). In chloramination systems, HAA5 formation is limited. In fact, ammonium chloride is added as a quenching agent in HAA5 compliance samples in order to halt HAA5 formation prior to analysis (*Standard Methods for the Examination of Water and Wastewater*, latest edition). Therefore, little additional HAA5 formation occurs after chloramination to offset HAA5 biodegradation occurring in the distribution system.

Enhanced Coagulation Requirement of the Stage 1 and 2 DBPR

The enhanced coagulation requirement has been developed to promote optimization of coagulation processes in conventional surface water treatment systems as required to improve removal of organic DBP precursors. The focus of the SWTR is separate from that of the enhanced coagulation requirement, with the former directed toward optimizing particle removal and the latter toward optimizing removal of natural organic matter (DBP precursors). Both promote efforts by water utilities to properly control and optimize coagulation processes and reduce DBP formation.

The enhanced coagulation requirements require treatment plants to remove specific percentages of total organic carbon (TOC) based on their source water TOC and alkalinity levels. Facilities must meet the enhanced coagulation requirements unless they meet any of the following exemptions (USEPA Stage 1 DBPR Guidance):

1. The PWS's source water TOC level is <2.0 mg/L, calculated quarterly as a running annual average.

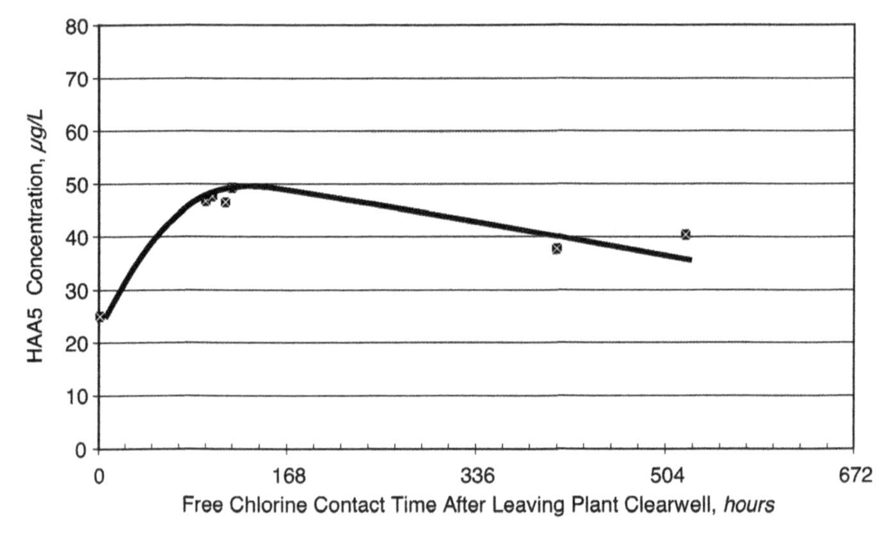

Figure 1-1 Formation and decay of HAA5 in a distribution system (time estimated by fluoride tracer test—T_{100})

2. The PWS's treated water TOC level is <2.0 mg/L, calculated quarterly as a running annual average.
3. The PWS's source water TOC level is <4.0 mg/L, calculated quarterly as a running annual average; the source-water alkalinity is >60 mg/L (as $CaCO_3$), calculated quarterly as a running annual average; and either the TTHM and HAA5 running annual averages are no greater than 0.040 mg/L and 0.030 mg/L, respectively; or the PWS has made a clear and irrevocable financial commitment to use technologies that will limit the levels of TTHMs and HAA5 to no more than 0.040 mg/L and 0.030 mg/L, respectively.
4. The PWS's TTHM and HAA5 running annual averages are no greater than 0.040 mg/L and 0.030 mg/L, respectively; and the PWS uses only chlorine for primary disinfection and maintenance of a residual in the distribution system.
5. The PWS's source water specific ultraviolet absorption at 254 nm (SUVA), prior to any treatment and measured monthly, is ≤2.0 L/mg-m, calculated quarterly as a running annual average.
6. The PWS's finished-water SUVA, measured monthly, is ≤2.0 L/mg-m, calculated quarterly as a running annual average.

Additionally, alternative compliance criteria for softening systems include the following:

7. Softening that results in lowering the treated water alkalinity to <60 mg/L (as $CaCO_3$), measured monthly and calculated quarterly as a running annual average.

8. Softening that results in removing at least 10 mg/L of magnesium hardness (as $CaCO_3$), measured monthly and calculated quarterly as a running annual average.

Utilities that cannot meet these avoidance criteria should know their enhanced coagulation endpoint, identified as the coagulant dosage and/or pH value that, when achieved, no longer produces significant TOC reduction. Specifically, when the source water TOC is not reduced by at least 0.3 mg/L with an incremental dosage increase of 10 mg/L alum (or equivalent ferric salt) and the pH value of the source reaches a value listed in Table 1-3, the enhanced coagulation endpoint has been reached.

If a utility is not exempt, a number of steps have to be evaluated relating to TOC removal, alkalinity of source water, range of source water TOC, required TOC removal for given source water characteristics, and several other factors.

Bromate

The bromate MCL from the Stage 1 DBPR remained at 0.010 mg/L for the Stage 2 DBPR. Bromate can be present in systems using ozone that have bromide present at the ozone application point. Bromate is also potentially formed during manufacture and storage of sodium hypochlorite. Consequently, systems using ozone for oxidation or disinfection are required to monitor once a month at distribution system entry point for bromate, but systems without ozone are not required to perform this monitoring. Systems that use ozone and that also add sodium hypochlorite will need to closely monitor the quality of these sodium hypochlorite products for bromate content.

Chlorite

Similar to bromate, chlorite monitoring is required only for systems using chlorine dioxide as an oxidant or disinfectant. Chlorite is a degradation product of chlorine dioxide. Chlorate is also a degradation product of chlorine dioxide but is not currently regulated. Chlorite and chlorate are potential degradation products of sodium hypochlorite, but systems using sodium hypochlorite are not required to monitor for chlorite unless they also use chlorine dioxide.

Table 1-3 Target pH values for coagulation when TOC removal rates are not sufficient

Alkalinity	pH Value
0–60	5.5
>60–120	6.3
>120–240	7.0
>240	7.5

Monitoring for chlorite is more complicated than for bromate because chlorine dioxide will degrade and chlorite formation will increase over time. Therefore, chlorite monitoring requirements include daily monitoring at the distribution system entry point and monthly samples at three locations in the distribution system (first customer, average residence time, maximum residence time). Unlike the health risks for bromate, TTHM, and HAA5, the risk for chlorite requires compliance based on average of the three chlorite-monitoring locations each month. The Stage 1 DBPR MCL for chlorite is 1.0 mg/L.

Residual Disinfectants

The maximum residual disinfectant level (MRDL) for combined or total chlorine is 4.0 mg/L as Cl_2. These are based on the same data used to monitor minimum free and combined chlorine levels in the distribution system as required by the SWTR, using the same monitoring locations used for the TCR. Chlorine dioxide residual also has an MRDL of 0.8 mg/L as ClO_2, based on daily samples at the treatment plant.

Surface Water Treatment Rule

IESWTR and LT1ESWTR

The goal of the IESWTR is to limit human exposure with harmful organisms, including *Cryptosporidium*, by promoting achievement of particle and turbidity removal targets for surface water treatment systems. Among IESWTR requirements that apply to surface water treatment plants are the following:

- Combined filter effluent turbidity must be ≤0.3 ntu for 95 percent of samples collected each month, including none with >1 ntu (compliance based on combined filter effluent samples collected at four-hour intervals during entire month).
- Utility must monitor each individual filter for turbidity at 15-minute intervals and must report results, including a filter profile (graphical representation of filter performance), if either of the following two conditions are met: (1) turbidity in any filter for two consecutive 15-minute intervals exceeds 1 ntu or (2) turbidity during first four hours of a given filter run exceeds 0.5 ntu for two consecutive 15-minute samples. Results must be reported within 10 days of the end of the month.
- Any newly constructed finished water reservoirs must include covers to keep out dust, debris, birds, etc.
- Utility must complete sanitary survey every three years. Existing surveys conducted after December 1995 can be used if they meet minimum requirements. Variances can be granted to decrease frequency to five years. The IESWTR explicitly requires that sanitary surveys include efforts to evaluate and control *Cryptosporidium*, in addition to other target organisms.

- Systems where the average of quarterly TTHM or HAA5 values exceeds 64 and 48 µg/L, respectively, need to complete disinfection profiling and benchmarking. *Profiling* involves determination of $C \times T$ values for each segment of treatment plant (see later discussion). *Benchmarking* involves determining lowest monthly average during 12-month monitoring of *Giardia* and virus inactivation. This procedure is required for any systems that are considering a major change to their disinfection practice. Consultation with the primacy agency is also required before any disinfection change.
- Turbidity monitoring records must be maintained for a minimum of three years.

Facilities in compliance with these requirements, chiefly the turbidity monitoring provisions, are designated by the IESWTR to have provided 2-log virus removal, 2.5-log *Giardia* removal, and 2-log *Cryptosporidium* removal. Literature and other information cited in the IESWTR final rule indicate that these credits are conservative, and most facilities meeting these requirements are probably achieving far greater levels of virus, *Giardia*, and *Cryptosporidium* removal than the minimum credits previously cited. The level of *Cryptosporidium* protection cited is sufficient to meet all requirements of the IESWTR, but the rule requires a total of 3.0 credits for *Giardia* and 4.0 credits for viruses. The additional credits (0.5-log for *Giardia* and 2-log for viruses) are required to be achieved by disinfection with free chlorine, chloramines, ozone, or chlorine dioxide by meeting *CT* requirements described later in this chapter.

Provisions of the IESWTR apply to large systems (>10,000 persons) using surface water sources. However, similar provisions are applied to smaller surface water systems (<10,000 persons), as outlined in the LT1ESWTR. The objectives of the LT1ESWTR and IESWTR are identical, though some of the compliance deadlines and other regulatory provisions are slightly different based on greater financial and personnel resources for larger systems.

Sanitary Surveys

Sanitary surveys are a requirement of the Interim Enhanced Surface Water Treatment Rule (IESWTR). A sanitary survey is "an onsite review of the water source, facilities, equipment, operation, and maintenance of the public water system for the purpose of evaluating the adequacy of such source, facilities, equipment, operation, and maintenance for producing and distributing safe drinking water." Surveys are usually performed by the state primacy agency and are required of all surface water systems and groundwater systems under the direct influence of surface water.

These surveys are typically divided into eight main sections, although some state primacy groups may have more.

1. Water sources
2. Water treatment process
3. Water supply pumps and pumping facilities

4. Storage facilities
5. Distribution systems
6. Monitoring, reporting, and data verification
7. Water system management and operations
8. Operator compliance with state requirements

Sanitary surveys are required on a periodic basis usually every three years. Surveys may be comprehensive or focused according to the regulatory agency requirements.

CT Requirements

Every water system that uses surface water as a source must meet treatment technique requirements for the removal and/or inactivation of *Giardia*, viruses, *Legionella*, and other bacteria. Because these pathogens are not easily identified in the laboratory on a routine basis, USEPA has set quality goals in lieu of MCLs in this instance. Meeting SWTR treatment technique goals demonstrates all or part of the required microbial protection, as previously noted, but additional protection is required through the use of approved disinfection treatment chemicals. The effectiveness of disinfection depends on the type of disinfectant chemical used, the residual concentration, the amount of time the disinfectant is in contact with the water, the water temperature, and, when chlorine is used, the pH of the water.

According to USEPA, a combination of the residual concentration, *C*, of a disinfectant (in milligrams per liter) multiplied by the contact time, *T* (in minutes), can be used as a measure of the disinfectant's effectiveness in killing or inactivating microorganisms. For water plant operators, this means that high residuals held for a short amount of time or low residuals held for a long period of time will produce similar results. Water plants are required to provide this computation daily, and it must always be higher than the required minimum value.

LT2ESWTR

The Long-Term 2 Enhanced SWTR (LT2ESWTR) supplements SWTR requirements contained in the IESWTR for large surface water systems (>10,000 persons) and the Long-Term 1 Enhanced SWTR (LT1ESWTR) for small systems (<10,000 persons). Details of the rules can be reviewed in the *Federal Register* or at the USEPA website (http://water.epa.gov/drink/index.cfm). One of the key elements of the LT2ESWTR was the use of *Cryptosporidium* monitoring results to classify surface water sources into one of four USEPA-defined risk levels called "bins." Facilities in the lowest bin (bin 1) are required to maintain compliance with the current IESWTR. Facilities in higher bins (bins 2 to 4) are required to either (1) provide additional *Cryptosporidium* protection from new facilities or programs not currently in use at a facility or (2) demonstrate greater *Cryptosporidium* protection capabilities of existing facilities and programs using a group of USEPA-approved treatment technologies, watershed programs, and demonstration studies, referred to collectively as the *Microbial Toolbox*.

Implementation of the LT2ESWTR was phased over many years according to system size. Four separate size categories were established (schedules 1–4 with 4 being the smallest <10,000 population) for implementing the rule. The rule for schedule-4 systems allows filtered supplies to perform initial monitoring for fecal coliform to determine if *Cryptosporidium* monitoring is required.

Filter Backwash Recycle Rule (FBRR)

The FBRR currently applies to systems of all sizes and is intended to help utilities minimize potential health risks associated with recycle, particularly associated with respect to *Giardia* and *Cryptosporidium*. Other contaminants of concern in the recycle stream include suspended solids (turbidity), dissolved metals (especially iron and manganese), and dissolved organic carbon. Plants that control recycle will also help minimize operational problems.

Prior to the FBRR, no USEPA regulation governed recycle. Regulations within the United States regarding recycle had been established by the states, if at all. State regulatory approaches varied from a requirement of equalization of two backwashes in Illinois to 80 percent solids removal prior to recycle and maintaining recycle flows at less than 10 percent of raw water flow in California. Virginia discourages recycling.

Key components of the FBRR include (1) recycle must reenter the treatment process *prior* to primary coagulant addition, (2) direct filtration plants must report their recycle practices to the state and may need to treat their recycle streams, and (3) a self-assessment must be done at those plants that use direct recycle (i.e., no separate equalization and/or treatment of recycle stream) and that operate fewer than 20 filters. The goal of the self-assessment is to determine if the design capacity of the plant is exceeded due to recycle practices.

GWR

The USEPA promulgated the final Ground Water Rule (GWR) in October 2006 to reduce the risk of exposure to fecal contamination that may be present in public water systems that use groundwater sources. The rule establishes a risk-targeted strategy to identify groundwater systems that are at high risk for fecal contamination. The GWR also specifies when corrective action (which may include disinfection) is required to protect consumers who receive water from groundwater systems from bacteria and viruses.

A sanitary survey is required, by the state primacy agency, at regular intervals depending on the condition of the water system as determined in the initial survey. Systems found to be at high risk for fecal contamination are required to provide 4-log inactivation of viruses. Increased monitoring for fecal contamination indicators may be required by the regulatory authority.

TCR and RTCR

The Total Coliform Rule (TCR) was finalized in 1989. The objective of the TCR is to promote routine surveillance of distribution system water quality to search for contamination from fecal matter and/or disease-causing bacteria. All points in a distribution system cannot be monitored, and complete absence of fecal matter and disease-causing bacteria cannot be ensured. The TCR is a regulatory approach for the implementation of monitoring programs sufficient to verify that public health is being protected as much as possible, as well as allowing utilities to identify any potential contamination problems in their distribution system. The rule requires monthly sampling at each distribution sampling point.

If a routine monthly sample is total coliform (TC) positive, the utility must determine fecal coliform (FC) or *Escherichia coli* (EC) in the same sample and also must perform verification monitoring by collecting a second sample and reanalyzing TC and FC/EC within 24 hours. The system is not in compliance if either of the following occurs: (1) if analysis and reanalysis of a given sampling location are TC positive (TC[+]) both times and FC[/EC+] at least one of these times or (2) if more than 5 percent of all monthly samples for a 12-month period are TC[+].

The TCR, and the Revised Total Coliform Rule (RTCR) that was finalized in 2013, impact all systems. The RTCR requires public water systems that are vulnerable to microbial contamination to identify and fix problems. The RTCR also established criteria for systems to qualify for and stay on reduced monitoring, thereby providing incentives for improved water system operation.

The RTCR also changed monitoring frequencies for some systems. It links monitoring frequency to water quality and system performance and provides criteria that well-operated small systems must meet to qualify and stay on reduced monitoring. It also requires increased monitoring for high-risk small systems with unacceptable compliance history and establishes some new monitoring requirements for seasonal systems such as state and national parks.

The RTCR rule further establishes an MCLG and an MCL for *E. coli* and eliminated the MCLG and MCL for total coliform, replacing it with a treatment technique for coliform that requires assessment and corrective action. The rule establishes an MCLG and an MCL of zero for *E. coli*, a more specific indicator of fecal contamination and potentially harmful pathogens than total coliform. USEPA has removed the MCLG and MCL of zero for total coliform. Many of the organisms detected by total coliform methods are not of fecal origin and do not have any direct public health implication.

Under the treatment technique for coliform, total coliform serves as an indicator of a potential pathway of contamination into the distribution system. A PWS that exceeds a specified frequency of total coliform occurrence must conduct an assessment to determine if any sanitary defects exist and, if found, correct them. In addition a PWS that incurs an *E. coli* MCL violation must conduct an assessment and correct any sanitary defects found.

The rule eliminated monthly public notification requirements based only on the presence of total coliforms. Total coliforms in the distribution system may indicate a potential pathway for contamination but in and of themselves do not indicate a health threat. Instead, the rule requires public notification when an *E. coli* MCL violation occurs, indicating a potential health threat, or when a PWS fails to conduct the required assessment and corrective action.

Lead and Copper Rule

Regulation

The objective of the Lead and Copper Rule (LCR) is to control corrosiveness of the finished water in drinking water distribution systems to limit the amount of lead (Pb) and copper (Cu) that may be leached from certain metal pipes and fittings in the distribution system. Of particular concern are pipes and fittings connecting the household tap to the distribution system service line at individual homes or businesses, especially because water can remain stagnant in these service lines for long periods of time, increasing the potential to leach Pb, Cu, and other metals. Although the utility is not responsible for maintaining and/or replacing these household connections, they are responsible for controlling pH and corrosiveness of the water delivered to the consumers.

Details of the LCR include the following:

- The LCR became effective Dec. 7, 1992.
- The action level for Pb is 0.015 mg/L and for Cu is 1.3 mg/L.
- A utility is in compliance at each sampling event (frequency discussed below) when <10 percent of the distribution system samples are above the action levels for Pb and Cu (i.e., 90th percentile value for sampling event must be below action level).
- Utilities found not to be in compliance must modify water treatment until they are in compliance. The term *action level* is used rather than *MCL* because noncompliance (i.e., exceeding an action level) triggers a need for modifications in treatment.

After identifying sampling locations and determining initial tap water Pb and Cu levels at each of these locations, utilities must also monitor other water quality parameters (WQPs) at these same locations as needed to monitor and evaluate corrosion control characteristics of treated water. The only exemptions from analysis of these WQPs are systems serving less than 50,000 people for which Pb and Cu levels in initial samples are below action levels.

Pb, Cu, and WQPs are initially collected at 6-month intervals, and then this frequency can be reduced if action levels are not exceeded and optimal water treatment is maintained. Systems that are in noncompliance and are performing additional corrosion-control activities must continue to monitor at six-month intervals, plus they must collect WQPs from distribution system entry points every two weeks.

Each utility must complete a survey and evaluate materials that comprise their distribution system, in addition to using other available information, to target homes that are at high risk for Pb/Cu contamination.

Revisions to the Lead and Copper Rule were enacted in 2007. These clarifications to the existing rule were made in seven areas:

- Minimum number of samples required
- Definitions for compliance and monitoring periods
- Reduced monitoring criteria
- Consumer notice of lead tap water monitoring results
- Advanced notification and approval of long-term treatment changes
- Public education requirements
- Reevaluation of lead service lines

Consult your local regulatory agency for those revisions that are applicable to your system.

Phase I, II, and V Contaminants

Regulations

The Phase I, II, and V regulations were finalized in 1989, 1992, and 1995, respectively, and include various inorganic and organic contaminants. Sampling and reporting frequency vary with constituent, though sampling is typically required once every three years after the initial sampling period. Variances or waivers are possible for a number of constituents based on analytical results and/or a vulnerability assessment.

Public Notification Rule

USEPA has implemented a regulation called the *Public Notification Rule.* This rule is separate from the Consumer Confidence Report (CCR) Rule. The Public Notification Rule includes requirements for reporting certain water quality monitoring violations and other water quality incidents, as well as requirements for the timing, distribution, and language of the public notices. For example, the Public Notification Rule includes requirements that some incidents be reported within 24 hours, others within 30 days, and others included as part of the annual CCR. Some of these reporting requirements are more stringent than those currently required by USEPA. The regulation also includes requirements regarding how notices are to be distributed/ broadcast (i.e., TV, radio, newspaper, hand delivery, regular mail, etc.), the format of the notices, the wording of certain items in the notice, and the need to include information in languages other than English.

Public notification according to the rule might include:

- Templates, or model notices, to be available for adaptation for certain potential incidents.
- Consolidated and updated lists of phone numbers and contacts for government (local, county, state), regulatory agencies, hospitals, radio and TV, newspapers, etc., that should be contacted per requirements of the Public Notification Rule.
- Checklists and flow diagrams outlining activities that would need to be completed for certain potential events outlined in the regulation.
- Identification of key personnel and what their roles and responsibilities would be to respond as required by the regulation.
- A plan to periodically review and update all lists, templates, and other aspects of a response plan every year or when/if the Public Notification Rule is modified by future federal or state regulations.

Unregulated Contaminant Monitoring Rule

The 1996 amendments to the SDWA require USEPA to establish criteria for a monitoring program for currently unregulated contaminants to generate data that USEPA can use to evaluate and prioritize contaminants that could potentially be regulated in the future. USEPA has developed three cycles of the Unregulated Contaminant Monitoring Rule (UCMR):

1. UMCR1 in 1999
2. UMCR2 in 2007
3. UCMR3 in 2012

Failing to (1) perform required sampling and analysis, (2) use the appropriate analytical procedures, or (3) report these results are violations of the UCMR. However, the numerical results of these analytical efforts cannot result in a violation because none of the constituents in the UCMR are currently regulated (i.e., no MCLs, action levels, or other standards apply).

Although the UCMR contaminants have no standards associated with them, the data from this monitoring will need to be reported in the annual CCR. Therefore, the CCR will need to address implications of any constituents found above detection limits. Reporting UCMR results in the CCR would also fulfill the notification requirements for "unregulated contaminants" included in the recently promulgated Public Notification Rule.

Note that the UCMR is an ongoing part of the regulatory development process that will be repeated every five years. Utilities will be performing similar mandatory sampling for a new list of constituents every five years.

The third Unregulated Contaminant Monitoring Rule (UCMR3) was signed by USEPA Administrator Lisa P. Jackson on April 16, 2012. As finalized, UCMR3 will require monitoring for 30 contaminants using USEPA and/or consensus organization analytical methods during 2013-2015. Together USEPA, states, laboratories, and public water systems (PWSs) will participate in UCMR3.

Operator Certification

Amendments to the 1996 SDWA required USEPA to develop national guidance for operator certification. The final rule was published on Feb. 5, 1999, and became effective on Feb. 5, 2001. State operator certification programs were required to address nine baseline standards, including operator qualifications, certification renewal, and program review. Indirect impacts of the rule on most water utilities include availability of Drinking Water State Revolving Fund (DWSRF) money and perhaps some slight modifications in paperwork/record-keeping requirements.

Arsenic

The MCL for arsenic was reduced from 50 μg/L to 10 μg/L in the *Federal Register* published on Jan. 22, 2001. This was the second time USEPA has established an MCL that was higher than the technically feasible level (3 μg/L), with the first being the uranium rule in 2000. The original SDWA required the MCL to be set as close to the health goal (zero for arsenic and all other suspected carcinogens) as technically feasible. Amendments to the SDWA allowed USEPA the discretion to set the MCL above the technically feasible level.

The final rule, including the revised MCL, became effective three years after the rule was published.

Radionuclides Rule

The Radionuclide Rule was published in December 2000. In the final rule, USEPA maintained the gross alpha MCL at 15 pCi/L MCL, 4 mrem/yr for beta emitters, 4 mrem/yr for photon emitters, and 5 pCi/L for combined radium 226 and 228 isotopes, and an MCL for uranium of 30 μg/L.

Analytical Methods

Each of the individual USEPA regulations contains their own information regarding analytical methods approved for compliance monitoring. These and other approved analytical methods are compiled in a final rule titled "Analytical Methods for Chemical and Microbiological Contaminants and Revisions to Laboratory Certification Requirements" published Dec. 1, 1999. These analytical methods were approved for compliance monitoring effective Jan. 3, 2000. The USEPA-approved methods include analytical procedures developed by USEPA, plus procedures developed by others that USEPA endorses, including specific procedures developed by the American Society for Testing and Materials (ASTM) and some specific procedures included in *Standard Methods for the Examination of Water and Wastewater*, published jointly by the American Public Health Association (APHA), AWWA, and the Water Environment Federation (WEF).

Currently, only approved analytical methods can be used for compliance monitoring. In the future, USEPA hopes to implement a performance-based measurement system that will allow utilities to use alternative screening methods instead of requiring only USEPA-approved reference methods. The 1996 SDWA Amendments require USEPA to review new analytical methods that may be used for the screening and analysis of regulated contaminants. After this review, USEPA may approve methods that may be more accurate or cost-effective than established methods for compliance monitoring. These screening methods are expected to provide flexibility in compliance monitoring and may be better and/or faster than existing analytical methods.

The approval of new drinking water analytical methods can be announced through an expedited process in the *Federal Register*. This allows laboratories and water systems more timely access to new alternative testing methods than the traditional rulemaking process. If alternate test procedures have the same or better performance of the approved methods, they can be considered for approval using the expedited process.

OPERATIONAL VIEWS OF CERTAIN CONTAMINANTS

Turbidity

Turbidity is the measure of the amount of particulate material in water. It is measured by detecting the amount of light scattered by the particles in a water sample. Turbidity is used as an indicator of water quality and as an indicator of the efficiency of certain removal processes such as coagulation and filtration. Adequate removal of turbidity is an important step in the process of removing pathogens. Although the turbidity measurement provides no information about the nature of the particles it measures, turbidity is considered very important and is thus regulated as a treatment technique. Optimization efforts, such as those found in the Partnership for Safe Water, are designed around the optimization of turbidity removal. High turbidity levels can make the disinfection process less efficient by creating higher disinfectant demand. Higher turbidities may also protect coliforms from disinfection by absorbing or encasing them.

Pathogen passage can be related to turbidity events (spiking) in the finished water. The minimization of the frequency and magnitude of spiking should be a top priority in water treatment. Operators should view any event of individual filter turbidity spiking as a potential for the passage of harmful pathogens into the water column. Each event should be analyzed for cause, and a plan for the elimination of the cause(s) should be implemented. Most filter-spiking events can be traced to operator involvement, unless plant inadequacies exist that make it impossible for water to be treated at times.

TOC

TOC is a composite measure of the organic content of the water. It is important for water suppliers to measure this contaminant because its presence and amount correlate to the production of DBPs. TOC can be removed in the coagulation/settling/filtration process, and removal efficiencies for this contaminant are regulated.

The challenge for treatment plant operators is to strike a balance between removal of TOC with traditional pH-lowering methods (more coagulant, more acid) and the need to maintain compliance levels with the LCR. Also, site-specific requirements for residuals disposal can be affected by the use of more coagulant for TOC removal. Especially difficult are the choices that operators consider when, although they are in compliance with DBP MCLs, they know they can achieve better results with more TOC removal through addition of more coagulant.

Waterborne Pathogens

Disease-causing organisms, called *pathogens*, include all of the problem-causing bacteria, viruses, protozoa, and algae. Well-known diseases caused by these organisms include typhoid fever, cholera, Legionnaires' disease, peptic ulcers, hepatitis, giardiasis, cryptosporidiosis, gastroenteritis—the list goes on. All of these pathogens can be found in natural surface water supplies and therefore can invade the treated water supply in sufficient numbers that will cause disease in the consumer. Even groundwater supplies may come under the influence of surface water and therefore be contaminated. Typically, pathogen exposure presents an acute risk rather than a chronic risk, and the detected presence of them or their surrogates will bring on boil-water notices. Given that some consumers are temporarily or permanently less immune than other consumers, pathogen removal is among the most important tasks confronting the water plant operator.

Bacteria, which are organisms that can cause severe illness in people, are usually minimized by conventional water treatment processes. They can exist in wide variations of temperature and water quality ranges, and some bacteria can pass through the water treatment plant and find their way to the consumer in small numbers. *Legionella* is such an example of an organism that can occur in the finished waters of systems employing full treatment. Of particular interest to water plant operators is the group of bacteria called *coliforms*. These organisms are used as indicator bacteria because they are easily cultured in the laboratory and can be an indicator of the presence of other more-difficult-to-culture bacteria.

Viruses, which are smaller organisms than bacteria, can also cause a number of debilitating diseases, including hepatitis and polio. They are difficult to culture, even in the most sophisticated laboratories. Fortunately, most viruses are removed or inactivated by conventional water treatment practices. The lime-softening process is very effective against viruses because of the high pH employed, and maintaining proper *CT* values improves disinfection of viruses.

Protozoa are larger than bacteria and can be very resistant to disinfection because of their ability to form spores or cysts that have tough outer casings. The notable protozoa that water plant operators deal with are *Giardia* and *Cryptosporidium*. These two organisms are found in nature and are typically removed in the filtration process. *Giardia* can be inactivated by sufficient $C \times T$ values employing chlorine, but *Cryptosporidium* is considered to be resistant to chlorine at levels far greater than those typically used in water treatment plants. The IESWTR regulates these two organisms (see text in this chapter) by requiring treatment techniques. When the treatment techniques are met, the utility is given "credit" for removal of *Cryptosporidium* and removal/inactivation of the *Giardia*. This credit must be achieved continually as the plant operates. Failure to meet the treatment technique results in boil-water notices.

Algae do not typically pose a health concern for humans, although some may produce neurotoxins and hepatoxins that can be of concern. The more frequent problems caused by algae are taste-and-odor episodes and filter-clogging episodes. Algae are difficult to coagulate and filter, and it is common for them to pass through filters in high numbers. Because of this, it is often more practical to prevent algae from growing than it is to remove them once they are in the plant.

Inorganic Chemical Contaminants

The inorganic chemical contaminants found in water supplies may be found as naturally occurring in the source or they may appear in the water through contact with the piping and storage components of the system. The chemical treatment process can also add to inorganic contamination. These chemicals in water can present chronic health problems at low levels and can also cause acute distress when ingested in high doses.

Many inorganic chemical contaminants have MCLs associated with them. Of notable exception are lead and copper, which carry action levels, as explained previously in the chapter. Most are regulated at the entry point to the distribution system, again with lead and copper being the exception. Iron and manganese are of special interest to the operator because of the staining problems they can create on fixtures. Removal of these nuisance contaminants is detailed in chapters 4 and 10. Nitrates are regulated with an MCL of 10 mg/L because they are believed to cause problems with the circulatory systems of infants.

Fluoride (see chapter 10) is an inorganic constituent with an interesting role in the water treatment business. Its presence in water is required by most states and communities (there are exceptions), but it is also a regulated contaminant. Adding to the confusion is the fact that fluoride carries with it both an MCL and an SMCL. The water plant operator, therefore, is confronted with the problem of adding a regulated contaminant to the water and then carefully controlling the amount that is added. The MCL for fluoride is 4.0 mg/L, and the SMCL is 2.0 mg/L. High levels of fluoride can cause mottling of the enamel of the teeth. Still higher levels can be poisonous to humans.

DBPs

Chlorine has played the primary role in disinfection of PWSs, but chlorine dioxide, ozone, and chloramines are also used. The use of these disinfectants can cause the formation of DBPs, as previously discussed. A number of these by-products are probably carcinogens and may cause other toxic effects. The rules governing the limitation of the DBPs set limits based on chronic exposure and tend to be regulated as averages for occurrence. The challenge for operators is to maintain adequate disinfection levels for microbiological protection while providing a water that is low in the by-products created in the disinfection process.

Chief among the DBPs (where chlorination is used) is the occurrence of THMs (see chapter 8), but the HAAs are also currently regulated. In time, other DBPs are expected to be regulated as health-effects data are brought forth.

When ozone is used as a primary disinfectant, there are potential concerns with bromate and formaldehyde production.

REGULATORY CHANGES

There is no doubt that regulations will continue to change based on results of research and a better understanding of the health consequences of various contaminants. Water plant operators must keep abreast of the latest developments and be prepared to adjust to these regulations. Water system operators will occasionally be presented with a dilemma when new research indicates that an unregulated contaminant may be harmful and it is found (perhaps in a very low concentration) in their water supply. What should the water utility tell its customers about this situation? Each utility should develop a communications plan and have it ready for this almost inevitable circumstance.

BIBLIOGRAPHY

AWWA. 2010. Principles and Practices of Water Supply Operations—*Water Treatment*, 4th ed. Denver, Colo.: American Water Works Association

Brown, R.A. 2004. Environmental Engineering and Technology Inc., Internal communication to author. December.

Edzwald, J.K., ed. 2011. *Water Quality and Treatment*, 6th ed. New York: American Water Works Association and McGraw-Hill.

Pizzi, N.G., and M. Rodgers. 2000. Testing Your Enhanced Coagulation Endpoint. *Opflow*, 26(2).

USEPA. Office of Groundwater and Drinking Water, website, http://water.epa.gov/drink/index.cfm

USEPA. 1993. *Guidance Manual for Compliance With the Filtration and Disinfection Requirements for Public Water Systems Using Surface Water Sources*, Appendix E. Science and Technology Branch, Criteria and Standards Division, Office of Drinking Water, USEPA.

USEPA. Guidance on Ground Water Rule (GWR). http://www.epa.gov/ogwdw/disinfection/gwr/pdfs/fs_gwr_finalrule.pdf

USEPA. Guidance on Total Coliform Rule. http://www.epa.gov/safewater/disinfection/tcr/pdfs/RTCR%20draft%20fact%20sheet%2061710.pdf

USEPA Final Implementation Guidance for the Stage 1 Disinfectants/Disinfection Byproducts Rule. http://water.epa.gov/lawsregs/rulesregs/sdwa/stage1/upload/s1dbprimplguid.pdf

USEPA LT2ESWTR Toolbox Guidance Manual. http://www.wqts.com/pdf/2003-01_LT2_ESWTR_Toolbox.pdf

USEPA. 1999. Guidance Manual for Conducting Sanitary Surveys of Public Water Systems; Surface Water and Ground Water Under the Direct Influence (GWUDI). Washington D.C: USEPA. June 22, 2009. http://www.epa.gov/OGWDW/mdbp/pdf/sansurv/sansurv.pdf

USEPA. 2004. Assessing Capacity Through Sanitary Surveys. Washington D.C.: USEPA: Drinking Water Academy. June 22, 2009. http://www.epa.gov/safewater/dwa/electronic/presentations/pwss/assesscapacity.pdf

Chapter 2

Source Water

The earth's water is ancient. All that is on the surface of the earth, and all that is hidden below ground, has existed since the beginning of time. This water moves in a constant cycle from the surface of the earth to the atmosphere by the processes of evaporation and transpiration and moves back to the earth by the processes of condensation and precipitation. Some of the water that falls back to earth percolates into the soil and becomes groundwater, while other water stays on the surface in lakes and streams. This constant recycling of water is called *the hydrologic cycle* (Figure 2-1). This cycle is important to water plant operators because the sources of supply that they draw from can only be replenished by the hydrologic cycle. Knowledge of this replenishment process provides a sense of how fragile the cycle really is and the need for conservation and stewardship of this precious resource.

Source water assessment is crucial when choosing treatment options that are designed into a water plant. The reliability of the processes that are ultimately chosen will be based on the historical water quality of the source when available. If the source is newly developed, great care must be taken to characterize it before the design concepts are realized. The resulting design must provide the water plant with the capability to provide consistent water quality that meets all applicable regulations and standards or goals.

Source water management techniques help operators because water management tends to minimize problems that may occur later in the treatment plant. It is often easier and more economical to protect the supply source from contamination than it is to deal with the contamination after it enters the plant. Source water protection, which is discussed later in this chapter, is the first barrier of a multiple-barrier approach to water quality.

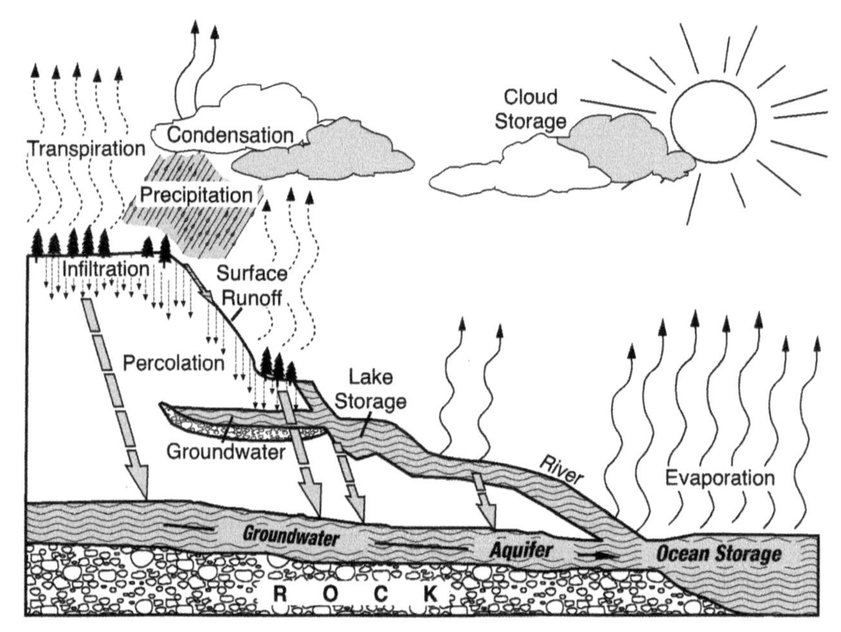

Figure 2-1 The hydrologic cycle

SOURCES OF WATER

Groundwater

Water that has percolated into the soil or seeped into the ground from streams and other surface sources is called *groundwater*. Water will continue to seep downward into soil until it reaches an impervious rock or clay formation. It then fills the voids of the soil or sand above the impervious cap and is referred to as an *aquifer*. If more water percolates through the soil, excess will accumulate in a zone of saturation, the top of which is called the *water table* (Figure 2-2). As long as water continues to fall into this area faster than it is taken away, the aquifer will yield a source of supply. Aquifers can be quite large, holding billions of gallons of underground water. Nevertheless, large cities and populations are capable of using this source so quickly that the water table can drop. In arid regions of the southwestern United States, this is a serious problem as the population migrates to these areas and makes more demands for water. Without adequate rainfall to replenish the aquifer, supply dwindles and conservation measures must be implemented.

Excessive withdrawal may impact the water quality of the aquifer. Examples are chronic seawater infiltration in the New York and New Jersey areas, as well as coastal Florida and California. Excessive amounts of iron and manganese can also be drawn into the influence of a well if it is operated at higher than normal rates.

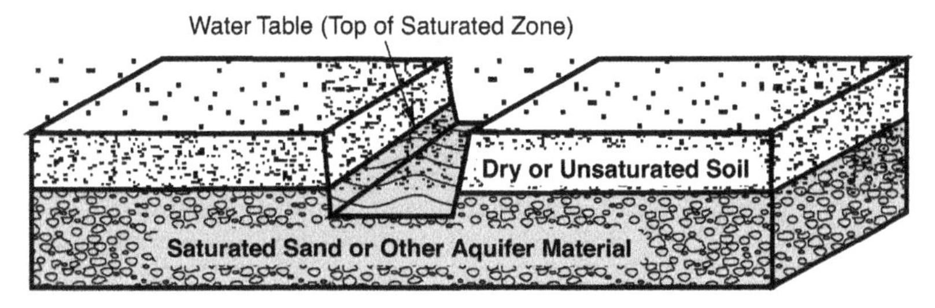

Figure 2-2 Schematic of a water table showing saturated zone

Groundwater occurs in two different kinds of aquifers: *unconfined and confined.* An unconfined aquifer, sometimes called a *water-table aquifer,* has an upper surface that is free to rise and fall with incoming and outgoing water (Figure 2-3). Wells constructed to draw from these sources can vary in their ability to yield a specific flow. A confined aquifer has an upper impermeable layer or a layer with low permeability. These aquifers are recharged at an elevation higher than the main portion of the aquifer. They are called *artesian aquifers* (Figure 2-4), and the water in them is usually under pressure. When these pressurized water tables intersect with the ground surface, they are called *springs.*

The nature and physical properties of the aquifer are important to the operator. The soil that holds the water will be porous to some degree. The more porous the soil, the more it will hold per unit volume. Also, the size of the soil particles will greatly influence the water's ability to move through the aquifer. Large-grained, sandy aquifers will allow water to move more freely than will smaller-sized sands, and pumping costs will be less. Some aquifers are made of limestone or fractured rock, and channels develop in them through which water can flow.

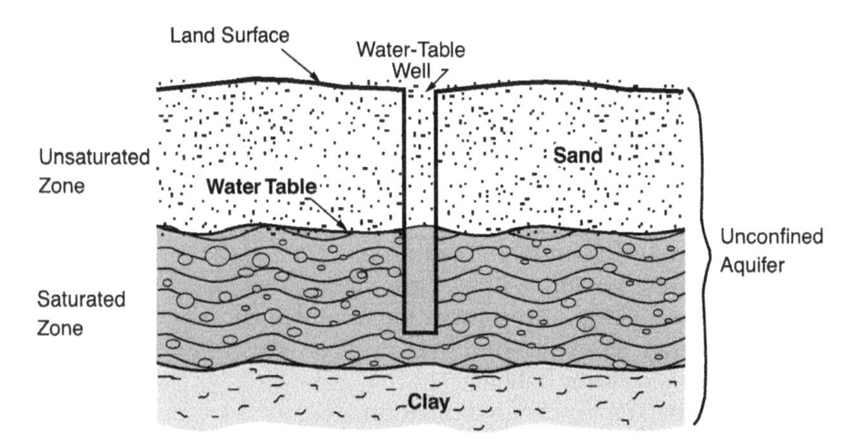

Figure 2-3 Schematic of an unconfined aquifer

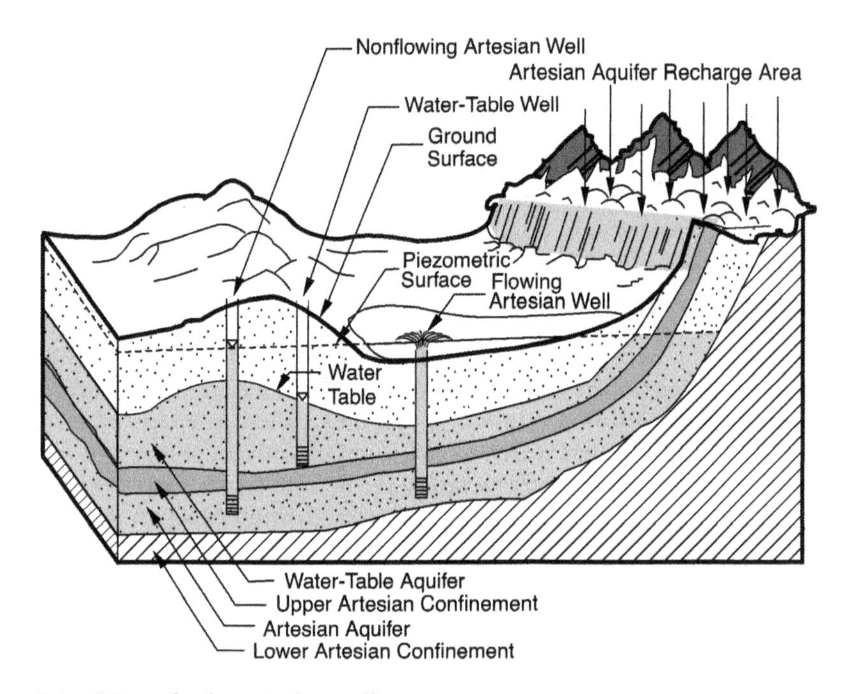

Figure 2-4 Schematic of an artesian aquifer

Though it is tempting to think of groundwater as free from microbial contaminants, it can become contaminated in many ways or come under the influence of contaminated surface waters. In 1989, the Surface Water Treatment Rule (SWTR) mandated that groundwater under the influence of surface water be treated similarly to surface water and that operators provide treatment techniques for microbiological contamination. In May 2000, the US Environmental Protection Agency (USEPA) proposed the Ground Water Rule, which targets the nation's groundwater supplies. It passed into legislation in October 2006. USEPA is determining, among other things, the amount of risk due to fecal contamination of these sources. Systems that are shown to be at risk for fecal contamination will be required to provide 4-log inactivation of viruses (99.99% removal). The USEPA is conducting surveys of the more than 147,000 groundwater systems in the United States and is looking at eight critical components to protect the public health. Corrective actions are required for any system with a significant deficiency or source water fecal contamination. The system must implement one or more of the following correction action options:

- correct all significant deficiencies,
- eliminate the source of contamination,
- provide an alternative source of water, or

- provide treatment that reliably achieves 99.99 percent (4-log) inactivation or removal of viruses.

The 1986 Amendments to the Safe Drinking Water Act (SDWA) called on each state to develop a wellhead protection program (WPP). This legislation established a national program to encourage states to develop systematic and comprehensive programs to protect public water supply wells fed by groundwater and well fields within their jurisdiction from contamination. (Certain WPP components are required to be implemented by states under the Source Water Assessment Program [SWAP], which was a part of the 1996 SDWA Amendments.) States encourage municipalities to participate and to exercise individual discretion in developing methods for protecting groundwater used for drinking water. The Source Water Protection Program (SWPP) incorporates surface water systems and is very similar to the WPP.

Watershed Sanitary Surveys

Sanitary surveys were discussed in chapter 1. However, some states require separate surveys that are focused on watershed issues. These are often incorporated into broader sanitary surveys but not always. Watershed surveys assess the adequacy of supply and examine the potential for contamination and may suggest protection strategies. Determining the capacity to sustain a viable, safe, water supply is a major outcome of these surveys.

Surface Water

Water found in lakes and streams is referred to as *surface water.* The availability and quality of this source of supply varies greatly by location. All surface supplies are assumed to be microbiologically or biologically contaminated to some degree and require some form of treatment prior to human consumption. The source for most surface supplies comes from precipitation—either rain or snow. Because the atmosphere contains a considerable amount of suspended impurities from volcanic activity and wind-blown soil, rain and snow become contaminated as they fall to earth. The amounts and types of impurities will vary by region.

When precipitation collects in lakes and streams, it becomes further contaminated as it travels over the terrain. It dissolves hardness-bearing minerals and can be subject to runoff conditions that may bring biological constituents to the water. As it passes over areas with organic content, the water collects natural organic matter (NOM), which is a precursor to disinfection by-products (DBPs).

Surface water can be soft or hard and may require different treatment for this characteristic. All surface water supplies require filtration and disinfection unless they can show certain avoidance criteria for filtration, as outlined in the SWTR and Interim Enhanced Surface Water Treatment Rule (see chapter 1). All surface supplies must be disinfected.

Water Impoundment Management

Some utilities construct impoundment reservoirs or use water from naturally occurring reservoirs that are fed from surface streams or that catch precipitation. Constructed reservoirs serve several useful purposes. They can store amounts of water in excess of daily demand and therefore act as a hedge against drought conditions. They can serve as pretreatment reservoirs when aeration systems or chemical treatment are added, or they may be used simply as large settling ponds prior to treatment. Reservoirs also allow the operator to draw water from nearby streams for storage. In this way, water can be pumped from the stream on days when water quality is excellent. When quality is poor, such as after a rainstorm or when there is an upstream spill, pumping can be discontinued if the reservoir is large enough to store sufficient water for treatment. When conditions return to normal, pumping from the stream can resume.

Reservoirs can also be a source of water quality problems because they are exposed to the elements. Sunlight penetration into the depths of the reservoir can promote algal growth, which causes taste-and-odor problems. Runoff from the ground can bring with it chemical and biological contaminants. Extreme wet and dry conditions affect water quality. For example, heavy precipitation can resuspend bottom sediments and increase turbidity. This can load the treatment plant with huge amounts of solids, making treatment expensive and management of residuals difficult. Drought conditions can cause stagnation and minimize the otherwise healthy dilution of contaminants because of low water levels.

The depth of a lake or reservoir is subject to two conditions that can impact source water quality. The location of the intake with respect to these depths is important. The first condition, known as the trophic cycle, involves three stages where nutrients and biological activity influence water quality. The *oligotrophic* stage is associated with low nutrients and limited algal production and biological activity. The *mesotrophic* stage involves a moderate amount of these constituents, and the *eutrophic* stage is rich in nutrients and biological activity. This eutrophic stage is associated with low oxygen levels, high turbidity and color, and NOM. Because oxygen is depleted, usually by algae, an anoxic condition can develop. This creates a reducing environment where insoluable iron and manganese bottom deposits can become soluble ions. If this water is brought into the plant, water quality problems can occur that can require treatment.

The second condition that can impact source water quality is *thermal stratification*. Layers of the lake or reservoir, called *limnion*, develop due to temperature differences. Oxygen depletion can take place within these layers and cause problems similar to those described previously. In colder climates, water at the top of a lake or reservoir can cool to around 39.4°F (4°C), a temperature at which water reaches its greatest density. This cooler, denser water sinks and forces the warmer water at the bottom to the top. This mixing can be so violent that the bottom of the lake or reservoir is riled, causing poor-quality, high-turbidity water to be brought into the plant. Operators refer to this phenomenon as *lake turnover* It can occur each fall and spring. Some operators have witnessed this phenomenon in uncovered sedimentation basins.

River bank filtration is considered groundwater under the direct influence of surface water. Treatment plants with this source of supply receive water of changing quality similar to the river but often the quality is more stable. The filtration of the soil either in the bottom or adjacent to the river provides beneficial pretreatment. In some situations, the degree of pretreatment has been assessed and the regulatory agency has granted some treatment credit to this process. This is site specific and at the discretion of the regulatory agency.

Direct river intakes often challenge water supply sources. Although there are advantages to drawing water directly from a river, such as contamination incidents that may flow downstream quickly beyond the intake, water quality can change rapidly compared to supplies from impoundments. Many possible contaminant situations may impact water quality: chemical spills from industrial discharges, river shipping accidents, floods, droughts, and surface runoff.

Operators who work at treatment plants with this source of supply need to be alert to these possibilities. Online, continuous monitoring is often used to provide warning of changes that may require adjustments to the water treatment system. Many treatment plants with direct river intakes have procedures in place to shut down production if certain conditions develop. Protecting the public water supply is of primary importance.

SOURCE WATER PROTECTION PROGRAM

A utility's watershed management program is the first barrier of protection against pathogen intrusion. Such a program can also protect drinking water quality by minimizing DBP precursors and protecting against algal growths that cause taste-and-odor problems. Deterioration of a watershed not only results in more costly treatment requirements at the plant but also can erode public confidence. If a consumer perceives that the watershed providing drinking water is not managed well, the consumer may seek alternative sources.

In an ideal setting, water utilities would own all of the land that comprises the watershed for their source. In fact, land ownership was the principal method of protecting source water in early protection programs. Currently, some utilities can still claim virtual ownership of their source's watershed; New York, Boston, Portland, Seattle, and San Francisco are among those. These utilities have avoided filtration requirements due in part to the high quality of their source water, which is a direct result of extensive land ownership and watershed management programs.

However, there is concern about microbial contaminants that are resistant to chlorination.By understanding the watershed and its sources of contamination, both active and potential, drinking water utilities can become advocates for improving their watershed and source water. Using tools such as those provided under the Clean Water Act (CWA) to achieve water quality goals, operators can develop a watershed protection program. In effect, Storm Water Pollution Prevention (SWPP) plans are simply an old idea that has received new attention to help guard against "new" threats. In general, a good watershed management program contains the following components:

- Identification and characterization of the source
- Characterization of potential impacts to the source
- Determination of the intake's vulnerability to contaminants
- Established protection goals
- Development of protection strategies
- Program implementation
- Program monitoring and evaluation

Many utilities have initiated SWPPs, including the Philadelphia Water Department. Philadelphia's experiences are described as an example SWPP. In colonial times, Philadelphia drew its water from wells, but by the mid-eighteenth century, the expanding city of nearly 30,000 had polluted this first source. Tanneries, which used unlined pits as part of the industrial process, and the large number of privies in the city had rendered the groundwater undrinkable. In 1799, after a series of yellow fever epidemics, Philadelphia decided to draw from the then "uncommonly pure waters of the Schuylkill River" by building the city's first waterworks at Centre Square (where City Hall stands today). In 1815, the city opened the Fairmount Waterworks to distribute water from the Schuylkill River throughout the city. As Philadelphia grew during the first half of the nineteenth century, its water supply was expanded to draw more water from the Schuylkill River to meet demand. The city then made efforts to ensure the purity of its source water. In addition to taking legal action to close upstream industries, Philadelphia began purchasing land along the banks of the Schuylkill River and Wissahickon Creek, a major tributary. This land was eventually set aside as a public park.

This watershed management experiment was successful for four decades, but upstream development eventually overwhelmed the city's program. With increased upstream industrial growth, particularly the coal mining industry, the Civil War, and population growth spurred by successive waves of immigrants, water quality continued to decline. By the 1880s and 1890s, Philadelphia experienced the worst typhoid and cholera epidemics of any American city as a result of deteriorated water quality. In 1900, the city decided to filter its water supply; by 1913, all of Philadelphia's drinking water was both filtered and chlorinated at five plants on the Schuylkill and Delaware rivers. An unfortunate by-product of this decision to invest in treatment plants, however, was a de-emphasis on source water protection. Not coincidentally, the period following construction of the water treatment plants saw considerable deterioration of the Schuylkill and Delaware rivers. This deterioration occurred in most major rivers across the country.

Eventually, Congress passed the CWA in 1972. In the 40 years since the passing of the CWA, most of the country's rivers and streams have been rehabilitated due to an emphasis on installing controls at municipal and industrial point source discharges, e.g., wastewater treatment technologies.

However, because of the threat from emerging pathogens such as *Cryptosporidium* and the realization that nonpoint sources such as urban and agricultural runoff now merit serious attention, Philadelphia and other cities around the country now realize that watershed management programs are critical to the long-term protection of public health. Watershed management programs represent the best opportunity for reaching beyond a city's limits and focusing on point sources to further enhance the source water. Consequently, in 1999, Philadelphia established a new SWPP to formalize and coordinate many ongoing research efforts, such as *Cryptosporidium*, geosmin, MIB (2-methylisoborneol), and bromide occurrence studies. The program also enabled the water department to have a single, cohesive voice when dealing with upstream stakeholders.

The Philadelphia Source Water Program began by collecting and analyzing data relating to existing water quality trends within the watershed. Additionally, the program began to actively complete total maximum daily loads (TMDLs) in the watershed by supplying sampling data and analyses, funding flow monitoring gauging stations, and providing technical assistance to the agencies conducting the TMDLs.

Another major milestone for the program was partnering with the Pennsylvania Department of Environmental Protection (DEP) to implement the 1996 SDWA-mandated SWAPs for Philadelphia's three treatment plants and 47 others within their watershed. DEP contracted with the Philadelphia Source Water Protection Program to perform SWAPs for 42 surface water intakes on the Schuylkill River. Additionally, DEP contracted with Philadelphia to conduct eight assessments on the tidal portion of the Delaware River, including Philadelphia's Baxter Water Treatment Plant.

The general approach (see Figure 2-5) for the assessments was to first identify all potential sources, if possible; develop and apply a screening procedure to narrow down the list to important sources; prioritize the important sources; and then follow up with data collection and reality checking of the top-priority sources.

The process for the Schuylkill River, for example, started with development of a database of more than 3,500 potential sources. The data was collected from various national databases such as the Permit Compliance System; Resource Conservation and Recovery Act Information System; Comprehensive Environmental Response, Compensation, and Liability Act Information System; the Toxic Release Inventory; Envirofacts; Right to Know Network; and Business Pro. Based on contaminant information for each source, the sources were assigned one or more of 10 pollutant categories.

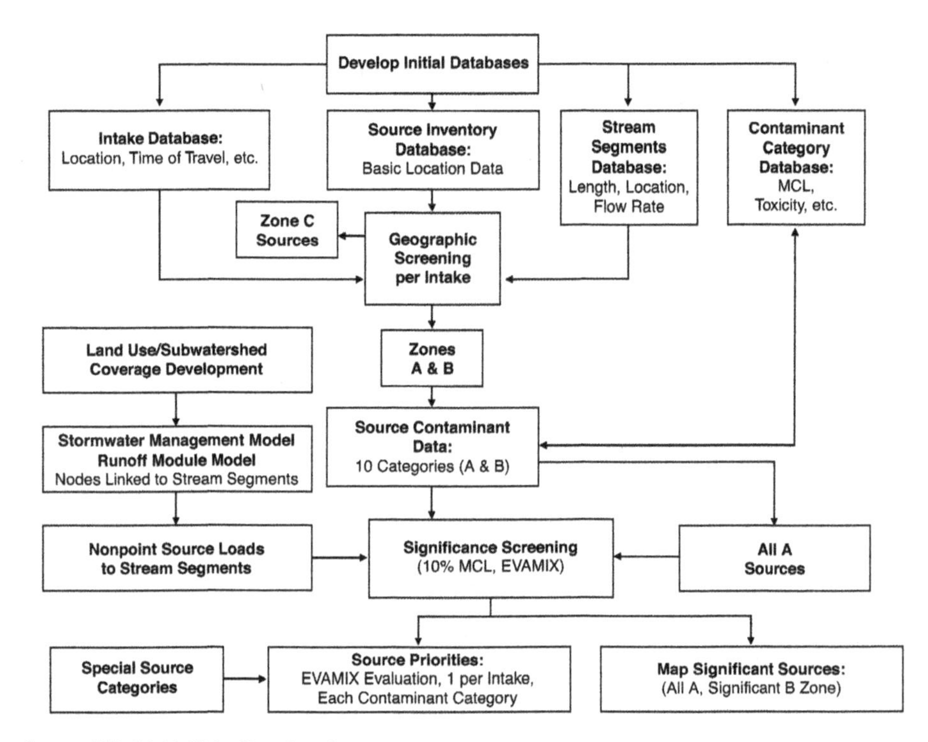

Source: Philadelphia Water Department.

Figure 2-5 Source prioritization flow diagram used by Philadelphia Water Department

Additionally, nonpoint sources are entered into the evaluation using a variation of the USEPA Storm Water Management Model, which utilizes land use coverage data for 300 subwatersheds to estimate annual loadings of contaminants as the result of stormwater runoff. Geographic screening is accomplished using delineated time-of-travel zones (A, B, or C) for each water plant intake. Finally, a computerized decision-support model (EVAMIX) is used to prioritize sources based on stakeholder-defined weighting criteria to determine top-priority source rankings. All high-ranking sources are flagged for follow-up data collection in later phases to verify results.

SOURCE WATER QUALITY CONSIDERATIONS

Many possible water quality issues confront water treatment facilities. Recognizing these challenges and being prepared to effectively adjust to them is the primary task for water plant operations personnel. It is the reason for extensive education and training. When the water quality changes and the treatment plant systems must be adjusted, experience and training are called on to meet the threat.

A treatment challenge is presented when the source water quality *changes*. A steady predictable water quality, even if it is not the greatest, can often receive effective treatment. The treatment system can be refined over time and excellent drinking water quality can be produced from even highly impaired waters.

Both surface and groundwater can change and potentially create a situation where water quality is at risk. Some of the water quality issues that may result from changes in source water quality are:

Bacteriological—e.g., fecal contamination
Biological—e.g., zebra mussels, algae
Chemical—e.g., radioactive substances, spills
Physical—e.g., temperature, pH
Agricultural—e.g., pesticides, fertilizer
Groundwater conditions—e.g., wellhead contamination
Lake stratification—e.g., taste and odor, iron or manganese

Direct river intakes can be very challenging water supply sources. Conditions can change quickly, so operators must be alert to these conditions. Treatment for known changes should be documented and plant adjustments rehearsed to ensure that acceptable drinking water quality is maintained.

Lake or other impoundments often stabilize water quality for lengthy periods. However, changes can occur, such as lake stratification and turnover. Another common cause of water quality changes in lakes is the growth of algae and other aquatic organisms. Some of these organisms can create disagreeable taste and odors. Others may contribute chemical by-products that can be toxic. Treatment of these conditions is best performed in the lake and controlled, if possible, to avoid these problems. Control of nutrients such as nitrogen and phosphorus compounds have been used to reduce algae and aquatic weed growth. In most instances, however, chemical control of algae is necessary. Copper sulfate is often the chemical used at a generic application rate of 1.1 lb/ac-ft. Also, strategies have been employed to reduce stratification by mixing or oxygen (air) application. These practices are often successful in reducing potential water quality problems but not eliminating them completely.

Groundwater sources can be contaminated by surface runoff entering an unprotected wellhead. This sort of incident can be serious because the well may not be monitored for this unforeseen occurrence. Disinfection strategies that include continuous monitoring and pacing chemical addition can be an effective defense when demand increases expectantly.

Understanding the characteristics of the water supply source and preparing treatment practices to combat quality changes are the best ways to maintain high quality water production. Treatment plant systems must be adjusted to meet changing water source conditions. Water plant operators should have proven plan in place to deal effectively with water source quality changes.

BIBLIOGRAPHY

American Water Works Association (AWWA). 2011. *Water Quality and Treatment*, 6th ed. James K. Edzwald, ed. New York: McGraw-Hill.

AWWA. 2010. *Algae Source to Treatment (M57), Manual of Water Supply Practice.* Denver, Colo.: American Water Works Association.

Pickel, Michael. 2001. Philadelphia Water Department, Internal communication to author. July.

Principles and Practices of Water Supply Operations—*Source Water*, 4th ed. Denver, Colorado: American Water Works Association (2010)

Chapter 3

Well Design and Operation

Exploration and site selection are important issues when developing well water sources because each well system is unique. Optimal production and use of a well source are possible only by planning in detail, properly testing and evaluating the data gathered, properly selecting pumping equipment, and properly operating the well field.

A well should be located so as to produce the maximum yield possible while still being protected as much as possible from contamination. State and federal geographical databases, which include information about water quality and availability, should be surveyed so that pertinent data can be examined for site location. Other information sources include landowners in the surrounding area and local commercial well-drilling contractors. Commonsense examination of the topography can lead to discovery of a well water source. It is likely to be present under valleys rather than under hills and it may exist where greenery is more lush and where streams or springs abound. It may be necessary to use exploration methods such as seismic testing, computer modeling, and test wells to confirm suspicions that water may be present.

The USEPA has implemented the Groundwater Rule, which requires proof of disinfection (for many groundwater systems) on a daily basis for the water drawn from the well prior to distribution to customers. Well-water system operators must comply with the $C \times T$ requirements that surface water systems have used for years. Some groundwater supplies have been granted a waiver from the filtration and disinfection requirements. These systems must monitor for certain contaminants to continue to qualify for the waiver.

TYPES AND OPERATION OF WELLS

A well functions in a way similar to the way a dam or reservoir captures surface water. An aquifer stores water, while a well provides a way to access the reserve of water. The well pump is not unlike the dam gate. It controls the flow of water from the reservoir. All wells work similarly in theory, but special considerations and differences are found from one well type to another. The following is a list of well types.

Dug Well

A dug well is capable of providing large amounts of water from shallow groundwater sources. They are literally dug by hand with pick and shovel; larger wells are dug with a backhoe. After excavation, the hole is lined with cement if the soil is unstable. Because they are large in diameter, dug wells are more difficult to protect from surface contamination.

Bored Well

With suitable soils, a well can be developed quickly by boring with an auger. These wells are limited to a 3-ft (1-m) diameter and may penetrate 25 to 60 ft (8 to 19 m). Bored wells are not common in public water supplies.

Driven Well

Driven wells consist of a pointed well screen, which is called a *drive point*, and lengths of pipe, which are attached to the point. Driven wells usually do not exceed 4 in. (10 cm) in diameter, nor do they probe deeper than 40 ft (13 m). For these reasons, they are not often used for public water supplies.

Jetted Well

Where soil permits, a special cutting tool that uses high-pressure water pumped through a cutting knife is used to "jet" a well hole. The casing is attached to the jet and is laid in place as the unit cuts into the ground. This system is not often used for public water supplies.

Drilled Well

Drilling of a well is the most common type of well development method for the public water industry. The method is versatile, can accommodate extreme depths, and can produce diameters that are commonly 4 ft (1.3 m) or more. Drilling methods include

- Cable tool method
- Rotary hydraulic method
- Reverse-circulation rotary method
- California method
- Rotary-air method
- Down-the-hole hammer method

Radial Well

Radial wells are special service wells that are commonly used near the shore of a lake or river to obtain large amounts of relatively good-quality water. Water quality can improve as the water travels through the sand or gravel bed in the stream and into the well. There is an interest in these wells because utilities may receive some removal credits for contaminants through the system. With this "riverbank filtration" credit, a utility may be able to avoid other, more expensive treatment technologies.

PARTS OF A WELL

Figure 3-1 shows the parts of a typical well. The well is equipped with a sanitary seal that prevents contamination from entering the well casing. The seal has openings into the well for discharge piping, air vents, and pump controls. Air vents let air in as the well level drops during operation. The well casing is a liner placed into the borehole to keep the walls from caving in. Grouting is cement that is pumped into the space between the drilled hole and the casing to protect the casing and prevent water from traveling along the casing from the surface or between aquifers. The well may have a screen if the source is in sandy or gravelly soil.

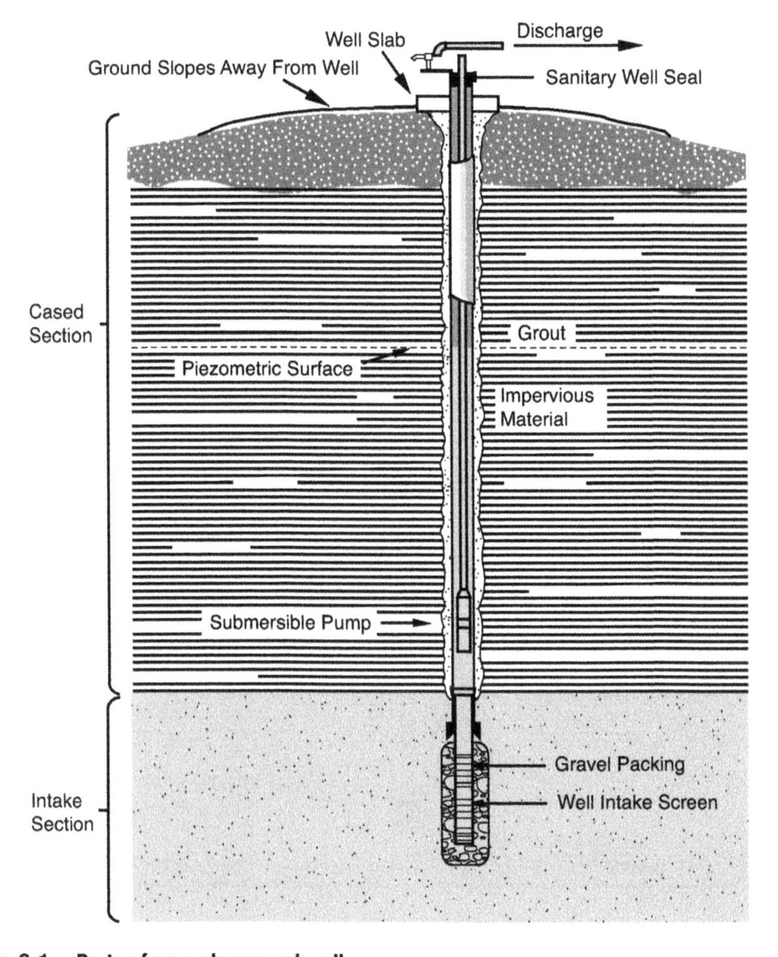

Figure 3-1 Parts of an underground well

OPERATIONS AND MAINTENANCE

Operators of well systems must be familiar with special terminology. These terms are listed below.

Static Level

The static level is the water surface in the well when no water is being pumped or taken from the aquifer. It is usually measured as the distance from the ground surface to the water surface. This measurement is used to monitor changes in the water table and should be taken and recorded frequently.

Pumping Level

The water level in a well usually drops when the pump is operating and water is being removed. When the water stabilizes at this lower level, it is measured as the pumping level. It is important that the pump intake be located below this level.

Drawdown

Drawdown is defined as the drop in water level between the static level and the pumping level (Figure 3-2).

Cone of Depression

In an unconfined aquifer, the flow of water in the aquifer to the well comes from all directions surrounding the well. The water surface in the aquifer takes on the shape of an inverted cone known as the *cone of depression* (Figure 3-2).

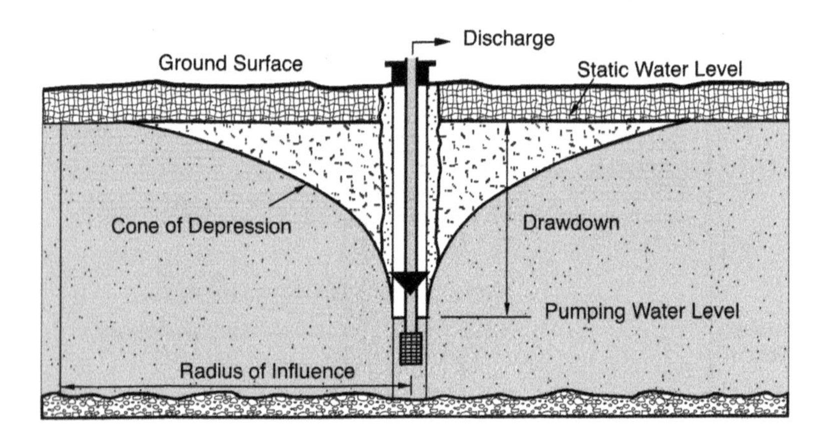

Figure 3-2 Drawdown in a well

Radius of Influence

The zone affected by drawdown extends outward from the well at a distance that is dependent on the porosity of the soil (Figure 3-2). The steepness of the cone corresponds to a radius of influence that extends horizontally outward from the well. Water will move easily through an aquifer composed of very sandy soil, and the cone will be almost flat. A more impervious soil will create a flow that deepens the cone and extends it out further. This is important to operators since wells can be situated near each other. If the radius or zone of influence of two wells overlaps, the wells will affect each other's yield.

Well Yield

The yield of a well is the rate of water withdrawal that a well can supply over a long period of time. Yield is measured in gallons per minute for small wells and in cubic feet per second for larger wells. When the pump rate from the aquifer exceeds the water recharge over time, the drawdown will gradually extend to greater depths, and the safe well yield will be reduced. This can allow the pump to suck air, which can potentially damage the unit. Operators then either lower the pump itself or operate the pump for shorter periods of time, allowing the well to rest and recover.

Recovery Time

Recovery time is the time that a well needs to rest in order to come back to safe yield.

Specific Capacity

Operation of a well system requires testing for specific capacity, which is a measure of the well yield per unit of drawdown. The formula for specific capacity is

$$\text{specific capacity} = \text{well yield} \div \text{drawdown} \quad \textbf{(Eq. 3-1)}$$

This example (see Figure 3-2) illustrates use of the formula. If well yield is 200 gpm (0.44 ft^3/sec) and the drawdown is 25 ft (8 m), the specific capacity is 200 ÷ 25 = 8 gpm/ft (1 ft^3/min/ft) of drawdown.

Surging

Fine particles that accumulate on the well face and reduce efficiency must be removed. This is done using a plunger, or surge block, that is moved up and down the well casing; the process is called *surging*. This maintenance procedure is performed periodically to ensure well efficiency. The process allows for the fines to fall to the bottom of the well, where they are removed with a bailer. Some well operators use air for surging. High-rate pumping coincident with surging may also be used to improve efficiency during the drilling process. High-rate pumping alone may also be used to clean out the casing in a rock aquifer.

These parameters should be tested and monitored at the same time each month so that comparisons can be made. Example operating data and calculations are shown in Table 3-1 and Figure 3-3. This facility is a precipitative softening plant that uses lime to treat a design flow of 3.7 mgd (5.7 m^3/sec). The plant normally operates 15 hours per day and treats about 1.6 mgd (2.5 m^3/sec). The source for the plant is well water drawn from 10 wells located within 6 mi of the plant.

Note that in Figure 3-3, the graphic shows that the September static level for well 27 was 65 ft, and the pumping level was 76 ft. This is a drawdown of 11 ft. An examination of Table 3-1 shows that the yield for well 27 was 400 gpm. The specific capacity can then be calculated (using Eq. 3-1) as 400 gpm ÷ 11 ft = 36.4 gpm/ft of drawdown. Note that although static levels have fluctuated due to recharge rates for well 27, the drawdown and specific yield levels are consistent. This is an indication that the well is performing properly.

Table 3-1 Orrville, Ohio, water treatment plant well performance monthly data bench sheet (partial)

	Well Number				
	27	**57**	**63**	**75**	**79**
Hour Readings	34,368.1	3,788.9	44,873.1	1,693.5	2,557.1
	33,957.9	3,746.1	44,546.9	1,633.1	2,189.2
	410.2	42.8	326.2	60.4	367.9
	Well Number				
	80	**83**	**84**	**85**	**86**
Hour Readings	1,154.1	3,569.3	68,104.4	375.8	32,360.8
	764.0	3,177.7	67,698.3	375.8	31,970.8
Total Hours	390.1	391.6	406.1	0.0	390.0

	Line Press	On	Off	gpm × 60 × hours = Flow			Well Flow Meter
27	95	94	66	400	410.2	9,844,800	7,998,000
57	—	40	22	280	42.8	719,040	565,000
63	44	56	44	470	327	9,221,400	8,107,000
75	52	31	22	290	60.4	1,050,960	1,235,000
79	96	206	166	340	367.9	7,505,160	7,589,000
80	90	138	116	480	390.1	11,234,880	10,001,000
83	80	170	116	500	391.6	11,748,000	12,114,000
84	120	64	53	260	406.1	6,335,160	4,644,000
85	140	68	42	100 (est)	375.8	2,254,800 (est)	954,000
86	68	156	131	240	390.0	5,616,000	5,958,000

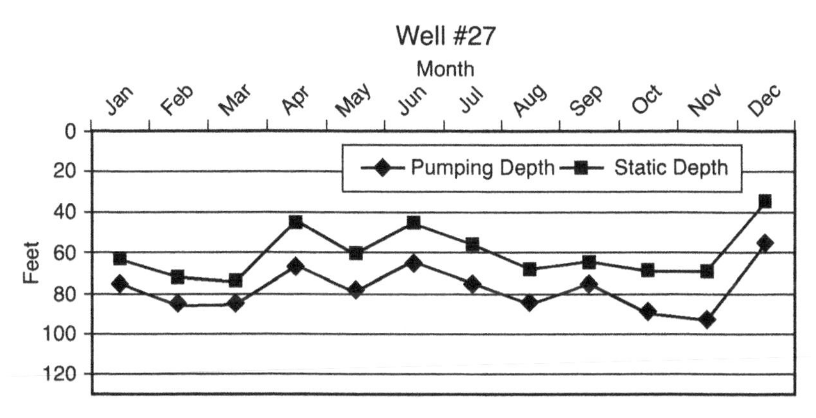

Figure 3-3 Year 2000 operating levels for well 27 at Orrville, Ohio, water treatment plant showing static and operating levels

Disinfection

Disinfection of wells is part of the startup and maintenance procedures that operators must follow. These techniques are discussed in chapter 8 and appendix A.

Operators should refer to AWWA Standard C654, Disinfection of Wells, for further reference.

Because the Ground Water Rule requires 4-log inactivation of viruses when using chemicals for disinfection, some groundwater systems will need to alter their disinfection processes. Groundwater system operators must understand the effect of these changes on distribution system water quality.

Table 3-2, which was published in the USEPA simultaneous compliance guidance, can be used as guidance to determine possible simultaneous compliance issues that may arise.

REASONS FOR WELL ABANDONMENT

When a well is no longer of any use, it must be abandoned properly. Wells need to be abandoned properly in order to

- Eliminate a physical hazard
- Prevent groundwater contamination
- Conserve the aquifer
- Prevent mixing of desirable and undesirable water between aquifers

In arid regions, well abandonment for aquifer preservation is not uncommon. Well sources that draw from aquifers bordering seawater are sometimes abandoned because continued use of the well would result in saltwater intrusion. In some types of soil, such as that found in Florida, wells have been abandoned because their use produces sinkholes.

Table 3-2 Possible simultaneous compliance issues

Rule or Regulation	Simultaneous Compliance Issue
Stage 1/Stage 2 Disinfection By-product Rules	Systems initiating chemical disinfection will need to begin distribution system monitoring.
Stage 1/Stage 2 Disinfection By-product Rules	Initiating or increasing chemical disinfection may result in disinfection by-product formation.
Stage 1/Stage 2 Disinfection By-product Rules	Systems using chemical disinfection must meet Stage 1/Stage 2 Disinfection By-product Rules' maximum contaminant levels in the distribution system.
Stage 1/Stage 2 Disinfection By-product Rules	Systems using chemical disinfection must monitor distribution residual levels and meet maximum residual disinfectant levels.
Lead and Copper Rules	Initiating chemical disinfection may result in changes in water quality parameters creating the potential for increased corrosivity in the distribution system and increased lead and copper contaminant levels at the tap.
Arsenic and Radionuclides Rules	Initiating chemical disinfection may require changes in corrosion control treatment for water quality parameters, which can impact the effectiveness of some treatments for arsenic and radionuclides contaminants.
Operator Certification	Initiating chemical disinfection may necessitate new operator certification requirements for the system.

BIBLIOGRAPHY

AWWA. 2010. Principles and Practices of Water Supply Operations—*Water Sources*, 4th ed. Denver, Colo.: American Water Works Association.

AWWA. 2010. Principles and Practices of Water Supply Operations—*Water Treatment*, 4th ed. Denver, Colo.: American Water Works Association.

AWWA. 2003. M21—*Groundwater*. Denver, Colo.: American Water Works Association.

Edzwald, J.K., ed. 2011. *Water Quality and Treatment*, 6th ed. New York: American Water Works Association and McGraw-Hill.

McGrew, J. 2001. Internal memo. October.

USEPA. 2007. Simultaneous Compliance Guidance Manual for the Long Term 2 and Stage 2 DBP Rules, U.S. EPA, Office of Water, EPA 815-R-07-017, March 2007.
http://www.epa.gov/safewater/disinfection/stage2/compliance.html

Chapter 4

Pretreatment

Pretreatment usually refers to treatment that occurs after withdrawing water from the source but before it enters the first (usually rapid mixing) process included in the water treatment plant. Pretreatment, or preliminary treatment, is often used by water utilities when the source water contains large debris such as sticks and leaves that might damage pumps or other process equipment. Pretreatment is also used when source water contains large amounts of grit, sand, or turbidity. Also, water containing some dissolved minerals or other impaired water quality conditions may benefit from some type of preliminary treatment. Pretreatment may be required continuously or only intermittently.

VARIATION IN SOURCE WATER QUALITY

As discussed in chapter 2, the quality of the source water will determine the design of the treatment plant and the methodology used to produce potable water. Water quality characteristics fall into four categories: physical, chemical, biological, and radiological. Some of the characteristics are considered to be contaminants and are undesirable. Operators base the quality of their source water on the concentrations of impurities found in it, and those concentrations vary by source and by characteristic.

One or more impurities will be present to some degree in all waters found in nature. Contaminants of concern include minerals such as iron and manganese, turbidity, color, pH, hardness, organic material, and total dissolved solids. Biological contaminants such as algae, zebra mussels, viruses, and *Cryptosporidium* are also a concern. The temperature and taste-and-odor characteristics of the source water will also affect water quality.

Source water quality differs from region to region and also from surface to groundwater supplies. Dissolved mineral concentrations of water vary widely and can contribute to many of the physical and chemical attributes of the water. Table 4-1 is a list of amounts of total dissolved minerals found in selected surface sources. Groundwater sources can also vary greatly in their mineral content.

Table 4-1 Typical mineral concentrations in surface water

Water Source	Total Dissolved Minerals, *mg/L*
Distilled	<1
Rain	10
Lake Tahoe	70
Suwannee River	150
Lake Michigan	170
Missouri River	360
Pecos River	2,600
Ocean	35,000
Brine well	125,000
Dead Sea	250,000

In general, groundwater sources tend to be more stable with respect to temperature, while surface supplies reflect temperature swings due to the local climate. Surface supplies in the northern United States may range in temperature from just above 0°C to 25°C or more. Consumers seem to prefer colder water, and industries that use water for cooling need colder water, too.

Tastes and odors may intensify at warmer temperatures. The temperature of water in lakes and reservoirs can vary with colder water generally found at the bottom. In winter in northern climates, this effect may be less noticeable.

Large bodies of water can develop "dead zones" where there is little or no oxygen. These zones may contain reduced (unoxidized) iron and manganese, and may be toxic to fish. If brought into the plant, the reduced forms of iron and manganese would exert an increased oxidation demand that may require special treatment. These dead zones can be natural or manmade. The Great Lakes water systems have seen an increase in the frequency and severity of dead zones. Operators should take note of rapid changes in water temperature, pH, alkalinity, and dissolved oxygen as indicators of dead zones.

Turbidity and color are optical properties of water that can vary greatly in surface supplies, in groundwaters under the influence of surface water, and in some regional groundwaters. Color can be inorganic or organic in nature, and its origin will dictate the type of treatment necessary. Color in surface water is derived from decomposed leaves and roots, other plant remains, and perhaps iron and manganese. Sewage can also be a source of color and is a health concern.

The pH of water, which is important to the treatment process, generally ranges from 6.5 to 8.5 S.U. (standard units) in surface waters but can vary from these values if pollution or acid mine runoff is present. Source water that varies too far from neutrality (pH = 7) will require some adjustment before distribution to customers. Groundwater pH generally ranges from 6.0

to 8.5. Hard water (see chapter 9) contains calcium and/or magnesium and can cause problems in plumbing systems, often requiring a utility to incorporate softening into its treatment scheme (see chapter 9). Table 4-2 shows hardness concentrations and the typical designations used to describe the water.

Organic constituents in water are a concern because they can be sources of disinfection by-products, as previously discussed in chapter 1. They also are a source of taste-and-odor problems. Estimates from various sources show that there may be more than 1,000 organic compounds that have been identified in drinking water supplies.

PROCESS DESIGN

Pretreatment processes, usually found downstream of the source, are designed to condition the water before it enters the treatment plant. A well-designed and well-operated pretreatment system will save the treatment plant capital and operating expenditures. Often, such a system can remove or minimize contaminants that occur only occasionally and that otherwise might require additional in-plant treatment capabilities that are only rarely used.

Screening

Screens and microstrainers are often used in the pretreatment process to remove gross debris and macrobiota (fishes, zebra mussels) before they can enter the treatment plant and damage pumps and appurtenances. As such, these are primarily physical processes.

Fixed screens, or bar screens, made from timbers or metal bars, may be located at the submerged intake. Screen openings are 1 to 3 in. (2.5 to 7.5 cm) in size and are designed to allow for velocities of less than 2 ft/sec (0.6 m/sec). Screens should be cleaned using water jets or air-scour mechanisms. Some screens are movable and can be raised to the surface for manual cleaning. Traveling screens (Figure 4-1), which are used at larger facilities, have multiple screens attached to a movable pulley. The pulley moves continuously, always bringing a new and cleaned portion of screen to the intake. Clogged screens move to the surface and are jetted. Microstrainers are often positioned downstream of intake screens or on source supplies where small particles and algae are abundant.

Table 4-2 Hardness concentrations with typical designations

Common Designation	Hardness, *mg/L as* $CaCO_3$
Soft	0–60
Moderately hard	61–120
Hard	121–180
Very hard	Greater than 180

Source: Johnson Screens Inc.

Figure 4-1 Traveling screens used to remove gross materials from the raw water supply

Oxidation

Source waters may be oxidated and aerated when the need arises; this need may be periodic or continuous. Iron and manganese, certain odors, carbon dioxide–laden waters, and water devoid of oxygen can be improved with preoxidation before the water comes into the plant. Refer to these chapters for specific information on oxidation process operation:

- Aeration—chapter 10
- Chlorine, chlorine dioxide, ozone, UV—chapter 8
- Potassium permanganate—chapter 10

Aeration is often the treatment of choice for iron removal. The oxidation state of iron can be elevated to +3, and this form combines with oxygen to become insoluble. The iron solids then settle or can be filtered more easily. Aeration is a very cost-efficient process for this purpose.

Manganese usually requires a more powerful oxidant to achieve conversion to the +4 form. This can be accomplished with free chlorine, chlorine dioxide, or ozone. Chlorine dioxide is an attractive choice because it does not form regulated organic disinfection by-products. The dosage must be controlled, however, because chlorite (a by-product) is regulated. About 2.5 mg/L of chlorine dioxide is required to remove 1.0 mg/L of manganese (site-specific conditions may alter this ratio).

Zebra and quagga mussels can be effectively controlled by the intermittent or continuous application of chemical oxidants. These small mollusks are invasive species that have caused serious problems in water supply pipelines. The organisms attached to the inside of transmission mains where they can disrupt flow and cause expensive increases in pumping requirements. Some water systems have applied chlorine, potassium permanganate, or chlorine dioxide at water intakes to control or eliminate these pests. Other methods of control include physical removal and UV disinfection near the point of water withdrawal.

Powdered Activated Carbon

To treat taste and odor or color in a water supply, it is sometimes useful to apply powdered activated carbon (PAC). The point of addition can be inside the water treatment plant, usually before sedimentation, but can also be near the supply source. Applying PAC near the source may have advantages because more time is available for adsorption of the substances of interest. This is especially true if presedimentation basins are available. Many times PAC requires substantial contact time to work effectively. Also the dosage needed may exceed the amount that could be applied in the treatment plant. PAC feed systems are similar to other dry chemical operations that include storage, dry feeders, slurry makeup, and solution feed components (see chapter 5).

Presettling

Operation of presedimentation basins usually requires regular removal of sand and grit; a sample program for testing the influent and effluent basins is a good idea. The frequency of testing may vary from season to season and may include an increase in sampling when water that is high in solids is taken in. As deposits accumulate in basins, the detention time lessens and more grit and residue enter the treatment train where damage can occur. Anaerobic conditions can also develop, causing treatment problems. Some packaged pretreatment units are cleaned automatically, while large uncovered reservoirs require periodic manual cleaning and drying.

Record-keeping practices are useful. The types and amounts of sedimentation, both influent and effluent, give a reliable indication of the unit's efficiency and can indicate a need for cleaning. Macrobiological sampling and analysis are also recommended, especially where zebra mussels and clams are present. The control of pests (like flies, gnats, or aquatic weeds) may require special pretreatment techniques such as preoxidation or physical scraping.

BIBLIOGRAPHY

AWWA. 2010. Principles and Practices of Water Supply Operations—*Water Treatment*, 4th ed. Denver, Colo.: American Water Works Association.

AWWA. 2004. M7—*Problem Organisms in Water: Identified and Treatment*, 3rd ed. Denver, Colo.: American Water Works Association.

Claudi, R., and G.L. Mackie. 2000. *The Practical Manual for Zebra Mussel Monitoring and Control.* U.S. Army Corps of Engineers Zebra Mussel Information System (ZMIS). Washington D.C.: CRC Press Inc. http://el.erdc.usace.army.mil/zebra/zmis/zmishelp.htm

Chapter 5

Coagulation and Flocculation

Most surface water supplies, and many groundwater supplies, contain particles that are too small to be removed by sedimentation without the use of chemicals. Large particles may be removed in presedimentation basins; however, small particles can take weeks or months to settle. Smaller suspended particles tend to resist settling because of their size and their natural repellant forces. These natural forces are excess electrical charges that tend to keep the particles in suspension because they help the particles repel each other. These small nonsettleable solids are referred to as *suspended solids*, *colloidal solids*, or *dissolved solids*, depending on their chemistry and origin. Typically, they carry a small negative electrical charge, which creates a repelling force much like the similar ends of a magnet. This force, called *zeta potential*, causes the particles to remain suspended.

Suspended particles can be biological organisms, bacteria, viruses, protozoa, organic material, or inorganic solids. Together, they give the water a cloudy appearance that is called *turbidity*. Visible turbidity is objectionable to consumers. It can also be accompanied by harmful biological agents and therefore must be removed. Turbidity can prevent pathogens from coming in contact with the applied disinfectant. The Safe Drinking Water Act requires that turbidity be removed to very low levels (see chapter 1) before distribution of water to customers.

To do this, water treatment plants normally apply chemicals to accelerate the removal of particles. These chemicals neutralize the zeta potential and allow the particles to bind together, become heaver, and settle. This process, known as *conventional treatment*, includes common concepts that operators should understand. Conventional treatment processes

- Are designed to perform at certain predetermined rates and under certain conditions.
- Can be optimized for performance, and so it follows that they can be compromised and overwhelmed.
- Taken together, form multiple barriers against the passage of pathogens into the distribution system.

The unit processes that are primarily responsible for conventional treatment particle removal include coagulation, flocculation, sedimentation, and filtration. The proper operation of these four processes will result in maximum removal of particulates from the source water and will produce an effluent with low turbidity. This chapter focuses on coagulation (rapid mixing here is considered as part of coagulation) and flocculation. Sedimentation and filtration are discussed in subsequent chapters.

COAGULATION

Coagulation is the addition and rapid mixing of chemicals, called *coagulants*, or *coagulant aids*, in water.

Chemical Coagulants

Common coagulants used in conventional water treatment include alum, ferric chloride, ferric sulfate, sodium aluminate, and various cationic polymers. The polymers are either used alone or in a form that has been combined with aluminum or iron salts (ferric salts). Both alum and ferric salts work by reacting with the water's natural alkalinity to form jelly-like particles of aluminum or iron hydroxide called *floc*. Sufficient amounts of alkalinity must be present for this to occur. If there is not enough alkalinity for the reaction to take place, operators may augment the naturally occurring alkalinity by adding caustic soda, soda ash, or lime (quick or hydrated).

Aluminum and iron coagulants put a large amount of positively charged ions into the water. These ions form positively charged flocs that begin to attract the negatively charged particles of color and turbidity. As the particles collide in the mixing area, they stick together and form increasingly larger particles.

The reactions of alum and ferric salts in water are governed by conditions such as mixing intensity, pH, temperature, alkalinity, and the nature and amount of turbidity. Alum works best in a pH range of 5.8 to 8.5, while ferric salts are effective over a wider range (4–11 under some circumstances). Both consume alkalinity as they form floc, causing the water's pH to drop. Ferric salts form a denser floc than does alum, and both form floc at a slower rate as the source water temperature decreases. Operators usually compensate for this by increasing the coagulant dosage. Initial application of the chemical is critical to floc formation, making it necessary to position the point of contact where mixing intensity is highest. Water that is highly colored with organic material requires a lower pH for coagulation; therefore, acid addition prior to coagulation may be necessary.

Coagulant Aids

Coagulant aids are used to improve the coagulation process and build stronger, more easily settled floc. They can also help to compensate for low temperatures and lower the required amount of primary coagulant. Coagulant aids may be used to raise alkalinity or increase the amount of solids in the water. The wastes produced during the coagulation process are called *residuals*, and they must be accounted for and disposed of properly. The residuals are a function of the amount of turbidity in the source water and the amount of coagulant and coagulant aid used to remove it. Refer to chapter 6 for more discussion of residuals.

COMMON COAGULANT AIDS

Activated Silica

Activated silica is a natural polymer that increases the rate of coagulation for alum and allows it to work over a wider pH range. It is typically used in amounts from 7 to 11 percent of the alum dose. The disadvantage of this aid is that the operator must activate it on-site. If the activation process is not carefully controlled, the silica will actually inhibit the floc formation and possibly clog filters.

Weighting Agents

If source waters are low in turbidity, it may be necessary to add particles to the water to enhance floc buildup. Weighting agents, such as bentonite clay, add particles to the water, which increases the probability of collisions between particles. Typically, dosages of 10 to 50 mg/L are used. This will add significantly to the amount of residuals produced.

Polyelectrolytes (polymers)

Polymers are very large molecules that carry highly charged ionic sites. Depending on the balance of the electrical charges they possess, they are classified as cationic, anionic, or nonionic:

- *Cationic polymers* possess an excess of positive charges and therefore are widely used to coagulate source waters with negatively charged particles. They are used as primary coagulants or as aids. Many water treatment plants use coagulants that are a prepared blend of alum or ferric salts with polymer. These blends help reduce the amount of metallic salts needed and therefore reduce pH depression. They also tend to form a tougher, more rapidly settling floc. They are very effective in flocculating living organisms such as algae. Cationic polymers are also commonly applied in very dilute solutions directly to waters just prior to filtration to improve filtration performance (filter aids).
- *Anionic polymers* possess an excess of negatively charged sites and are used for source waters or other process points in the plant where turbidity is positively charged. These compounds are often used at very low dosages as filter aids.

- *Nonionic polymers*, as the name implies, possess a balance of anionic and cationic sites and therefore have a neutral charge. They release both positive and negative charges into the water and can coagulate waters with varying zeta potentials. Low dosages are often used when these compounds are applied directly ahead of filtration (filter aids).

Anionic and nonionic polymers are typically suited as flocculation aids (fed just after cationic flocs are formed, e.g., after a minute or so of mixing) or for residuals thickening (in solids settling tanks or before application to a sludge press, sand bed, or centrifuge). These polymers are often formulated from regulated substances, i.e., acrylamide or epichlorohydrin, and are usually fed at low doses (less than 1 mg/L). Their use in drinking water treatment requires verification that they will be maintained at safe levels. Overdosing of polymers can lead to poor sedimentation and filtration.

COAGULATION CHEMICAL FEED SYSTEMS

Successful coagulation depends not only on the choice of coagulant but also on reliable feed equipment and proper mixing facilities. Additionally, operators must have a basic understanding of dosage control and process optimization.

Chemicals are stored as liquids or solids, depending on the amounts used and the size of the plant. Larger plants tend to use liquid products, which can be delivered in bulk. Many of the polymers that are used in smaller amounts can be handled in dry form, which is delivered in bags or drums. Operators usually prefer liquid chemical storage and handling because it creates less dust, fewer safety concerns, and fewer inventory problems. The choice of chemical type will have a bearing on the type of feed system used.

Dry Chemical Feed Systems

Two general types of dry chemical feeders are used in water treatment plants: volumetric and gravimetric. Both systems are equipped with mixers and a batch or mixing tank that provide a place for the chemical to mix with some "carry" water. This disperses the chemicals into solution or suspension prior to the application point. It is important that dry chemicals be thoroughly mixed before they are added to the treatment process.

Volumetric feeders (Figures 5-1 and 5-2) transfer a chosen volume of chemical from a hopper to a mixing tank in a unit time. They have adjustable settings that allow for dosage control. Volumetric feeders are not as accurate as gravimetric feeders but are less costly. Gravimetric feeders (Figure 5-3) measure out a predetermined weight of chemical per unit time.

Both volumetric and gravimetric systems can be operated automatically or manually. When operated in the automatic mode, they are often set to run from a signal of raw water flow or zeta potential or from other such controllers. Operators should understand that these automatic systems can fail and therefore do not guarantee trouble-free operation. Feeders should be checked

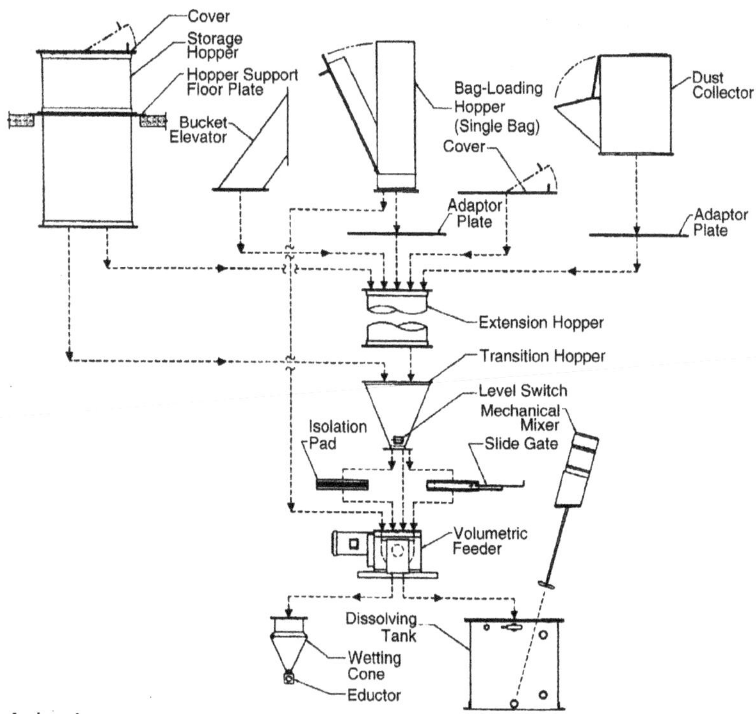

Source: Acrison Inc.

Figure 5-1 Schematic of a volumetric feeder with bag loader hopper

Source: Acrison Inc.

Figure 5-2 Volumetric feeder

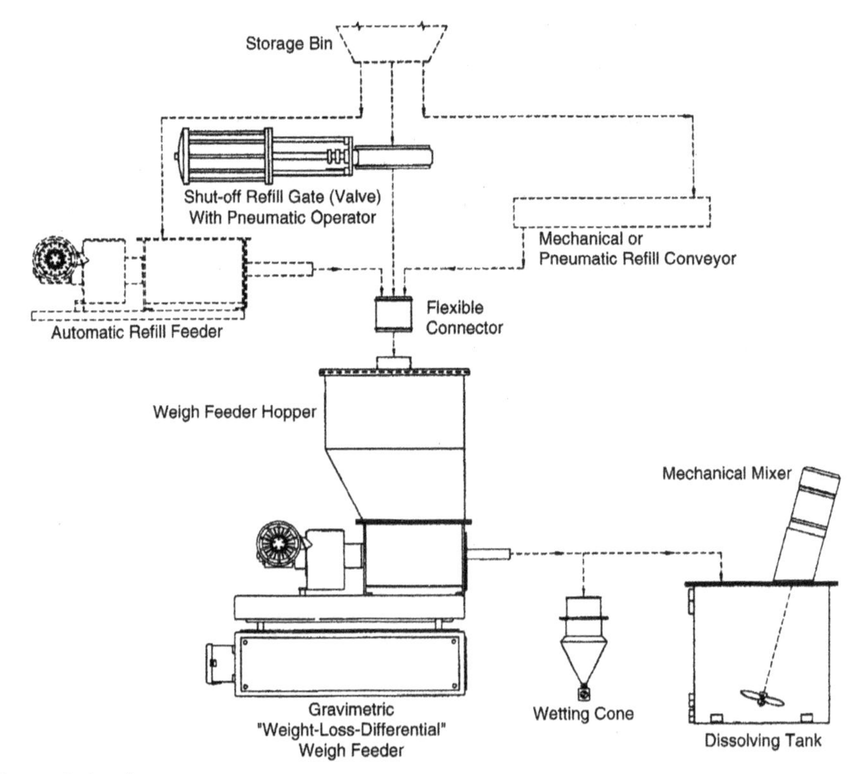

Source: Acrison Inc.

Figure 5-3 Schematic of a gravimetric feeder

systematically to confirm that they are doing what they have been programmed to do. Automatic controls can, however, provide a measure of reliability and free operators to perform other tasks. Also, automatically controlled feed systems allow for dependable operation of treatment plants from remote locations.

Solution Feed Systems

Liquid coagulants and coagulant aids are fed either full strength or as a dilution using solution feeders (Figure 5-4). When the chemicals are fed "neat" (directly from the package without dilution), they are commonly delivered through a metering pump (Figure 5-5). These positive-displacement units are usually variable-speed driven and can be paced from a signal of choice (usually raw water flow or zeta potential). Small solution feed systems offer the advantage of low maintenance and excellent, precise control and can be set up to feed directly from chemical drums that are delivered to the plant. For very small applications, peristaltic pumps are often used. These pumps have a roller mechanism that squeezes the coagulant through a tube or flexible hose, which must be replaced periodically. They are very accurate, even

Source: Cleveland, Ohio, Dlvision of Water.

Figure 5-4 Alum feed equipment used for liquid alum

Source: LMI Milton Roy.

Figure 5-5 Chemical feed pump

at low flows. Also, because no chemical actually touches the pump, there is no internal corrosion problem.

Solution feeders can be arranged with several tanks in series, one of which can be used as a dilution tank. With this setup, a higher-strength chemical can be purchased and dilutions made and fed as needed. This allows for less frequent purchases and perhaps a lower unit price for chemicals.

A solution feed tank can be set on a scale to accurately measure weight loss or the tank can be calibrated with markings at intervals so that volume-loss measurements can be recorded. Alternatively, the tanks can be equipped with sonic measurement devices that give accurate readouts of tank volume.

PROCESS OPERATIONS AND MAINTENANCE

Process control instrumentation, such as streaming current monitors, can help operators to produce a quality product. To achieve high-quality water, a water treatment plant must be operated in a consistently optimized manner and plant personnel should be able to

- Follow a written procedure (standard operating procedure [SOP]) that outlines and directs bench- or pilot-scale testing (jar testing, etc.) that determines proper chemical application dosages and sequences for their plant
- Take the knowledge gained in this determination and apply it to the plant full-scale

Calibrating Jar Tests for Dosage Control Evaluations

The details of how to perform jar tests are described in chapter 12 and in the AWWA manual M37—*Operational Control of Coagulation and Filtration Processes*. One of the most important steps for obtaining the most accurate jar tests is to ensure that the results mimic the actual results observed in the full-scale water treatment plant. Many operators lament that the jar tests "don't work" or that they are a "waste of time" because the results are not what is seen in the plant. They are correct. If the tests do not predict plant results, why do them? It is a proven fact that the results can be used to predict full-scale plant performance. However, the jar test procedures must be calibrated to yield these results.

The most important factor when calibrating a jar test is to get the hydraulic settings correct. The amount of time and the mixing intensity are critical to simulate plant conditions. Rapid mix, flocculation (time and mixing in each stage), and settling time are all important. Other factors to consider are temperature and light. Cold water in particular will affect the results. Some plants have constructed cold water baths around the jars to keep the water cold during the tests. Where algae are present, the effect of sunlight may also have an effect.

Calibrating jar test settings so that plant performance can be simulated takes time and patience. The following is a calibration process that can be used to lessen the frustration and ensure the best results.

1. Ensure current plant conditions and results
2. Utilize hydraulic settings estimating procedure for a starting point
3. Apply chemical dosages and hydraulic settings bracketing current conditions
4. Record best results settings for future reference
5. Perform calibration process when new conditions present themselves
6. Record new settings to give a complete calibration for all conditions
7. Use precalibrated jar test conditions when performing future tests

To start a calibration test, check all of the current plant operating conditions. Do not assume anything is correct. Check the flow rate, rapid mix conditions, chemical feed rates and dosages, flocculation mixer settings, number of flocculation basins in service, sedimentation basins and flow through each, turbidity and appearance of floc at each stage, other test results such as pH, and temperature. These values should be recorded for a comparison jar test.

The hydraulic settings estimator should be used to establish the jar test mixing and settling times. The settings will need to be changed one at a time until the results are similar to the full-scale plant.

Jar Test Hydraulic Settings Estimating Procedure

To estimate jar test hydraulic settings, proceed as follows:

- Rapid mixing G value and jar test setting—most plant rapid mixers provide a G value greater than can be duplicated in the jar tester, so mix for 30 sec at max of 300 rpm
- Flocculation of the plant usually is done in several stages. Calculate the time in each and get the G values from the plant design specifications. Set the jar tester to match. Detention time calculation is shown in appendix B. Jar tester G values for various rpm settings include

100 rpm	80 G
80	70
60	50
40	30
20	13

- Sedimentation time (min) is calculated for the plant. In the jar, the sample port is usually 10 cm from the water surface. Calculate the time in the plant sedimentation basin to settle 10 cm. Example: Surface overflow rate is 0.5 gpm/ft^2. What is the jar settling time for a sample 10 cm below the surface? 10 cm is 3.9 in.; 0.5 gal is 0.07 ft^3. So, 0.07 ft^3/min/ft^2 = 0.07 ft/min. or 0.8 in./min. It will then take about 5 min to settle 3.9 in. or 10 cm. Try to collect a sample in 5 min and compare to the plant settled water turbidity. Many plants find that 10 min works best.

Repeat these steps to develop settings for most treatment flows and situations. It is best to do the calibrations during a time when conditions are not changing. Do not rush. Constructing a set of calibrated jar test settings will be necessary to react to a water quality challenge. A calibrated jar test will provide good dosage settings. The dosage can be refined if the quality challenge continues. Consult the simplified jar test procedure in chapter 12 or the detailed procedure in M37—*Operational Control of Coagulation and Filtration Processes* (AWWA Manual of Practice) to determine the dilutions needed for the dosage calculations.

Coagulant Dosage Calculations

Plant operators use jar testing to evaluate various chemicals and dosages, on the bench scale, that will help them adapt to changing source water quality. (Jar testing is covered in depth in chapter 12.) However, the information gathered through jar testing is useless if the operator cannot take that information and apply it properly. This means that the operator must understand dosage calculations and how to apply the correct amount by adjusting the feed system.

Alum, the most widely used chemical in the coagulation process, has the formula $Al_2(SO_4)_3 \cdot 14.3H_2O$. As dry alum, it is 17 percent basis Al_2O_3. Each milligram per liter of alum added to water will consume approximately 0.5 mg/L alkalinity. Liquid alum, purchased and used in bulk quantities, is really dry alum diluted with water to about 48.8 percent, dry basis. Liquid alum weighs approximately 11 lb/gal at an average specific gravity of 1.325 (8.34 × 1.325). Liquid alum contains about 5.36 lb/gal dry-basis alum (48.8 percent of 11 lb/gal). With this information, the strength of alum, in milligrams per milliliter, can be calculated and used in dosage problems. The calculation is

$$5.36 \text{ lb dry alum/gal} \times 1 \text{ gal}/3.785 \text{ L} \times 453.6 \text{ g/lb} \cong 642 \text{ g/L, or } 642 \text{ mg/mL}$$

Therefore, each milliliter of liquid alum puts 642 mg dry-basis alum into the water. Dosage problems can now be solved, as in the following example.

Example: After performing a jar test, a water plant operator chooses a dosage of 24 mg/L dry-basis alum. The plant is operating at 1.6 mgd. How many milliliters per minute liquid alum must be fed?

1. *Convert 1.6 mgd to L/min.*

   1,600,000 gal/day × 3.785 L/gal × 1 day/1,440 min = 4,205.5 L/min water treated

2. *Calculate the amount of liquid alum needed.*

   24 mg/L × 4,205.5 L/min × mL alum/642 mg ≅ 160 mL/min

Alternative check on calculation:

$$160 \text{ mL/min} \times 1{,}440 \text{ min/day} \times 642 \text{ mg/mL} \times \text{lb}/453.6 \text{ g} = 326 \text{ lb dry alum/day}$$

$$326 \text{ lb dry alum/day} \div 1.6 \text{ mgd} \div 8.34 = 24 \text{ mg/L}$$

After performing the calculation and setting the feed equipment to deliver the needed amount of chemical, the operator must periodically check the equipment to determine if it is actually feeding the required amount. For liquid feed systems (described in the previous example), operators must have a way to capture or measure the amount of liquid fed per unit time, usually each hour. Dry feed systems require a method for measuring the weight of chemical fed per unit time (Figure 5-5). Both systems are normally equipped with such features. Liquid feed systems that use metering pumps are sometimes rigged with volumetric site tubes on the suction side of the pumps. Older volumetric feeders, such as the "rotodip" feeder, require the operator to capture the liquid as it is fed or to know the volume displaced from each dip. A chart for suggested feed rates of liquid alum is given in Table 5-1.

For raw water pump rates other than those shown, use the appropriate multiplier or extrapolate. Two examples follow:

Example 1: A water treatment plant operator is making a pump change that will create a total raw water flow of 2.5 mgd. The desired alum dosage is 20 mg/L. How many milliliters per minute of alum should be fed?

Answer: A 20-mg/L dose for 2 mgd requires 164 mL/min. A 20-mg/L dose for 3 mgd requires 246 mL/min. Extrapolating between the two shows that about 205 mL/min are needed.

Example 2: A jar test indicates that 17 mg/L of alum should be fed into a raw water flow of 80 mgd. How many milliliters per minute of alum are needed?

Answer: To treat 8 mgd with a 17-mg/L dose of alum, 557 mL/min are needed. Therefore, to treat 10 times as much flow, or 80 mgd, multiply 557 by 10. Thus, 5,570 mL/min are needed.

Table 5-1 Amount of liquid alum needed to achieve a desired dosage at a given pumpage

mg/L Dry-Basis Alum Desired	Flow Rate, *mgd*								
	1	2	3	4	5	6	7	8	9
	mL/min Needed	mL/min Needed	mL/min Needed	mL/min Needed	mL/min Needed	mL/min Needed	mL/min Needed	mL/min Needed	mL/min Needed
12	49	98	147	196	246	295	344	393	442
13	53	106	160	213	266	319	372	426	479
14	57	115	172	229	286	344	401	458	516
15	61	123	184	246	307	368	430	491	552
16	65	131	196	262	327	393	458	524	589
17	70	139	209	278	348	417	487	557	626
18	74	147	221	295	368	442	516	589	663
19	78	155	233	311	389	466	544	622	700
20	82	164	246	327	409	491	573	655	737
21	86	172	258	344	430	516	602	687	773
22	90	180	270	360	450	540	630	720	810
23	94	188	282	376	471	565	659	753	847
24	98	196	295	393	491	589	687	786	884
25	102	205	307	409	512	614	716	818	921

Note: Chart assumes that liquid alum is 642 mg/mL dry-basis alum.

To use the table:

1. Determine, through jar testing or some other means, the desired dosage of dry-basis alum.
2. Locate that dosage in the left column of the table.
3. Choose a treated water flow rate in the top row of the table.

The intersection of the column and the row shows the approximate amount of liquid alum needed, in milliliters per minute, to achieve the desired dosage.

Some water treatment plants use an alum–polymer blend as their primary coagulant. Normally, the dosage for this product is expressed as gallons per million gallons, or parts per million. This is based on the simple idea that 1 gal of product added to each million gallon of raw water is 1 ppm. This expression negates the requirement to know the exact proportions of alum and polymer in the coagulant.

Note that parts per million and milligrams per liter are not necessarily the same. Parts per million is a volume-per-volume relationship, while milligrams per liter is a mass-per-volume relationship. As a rule of thumb, a dosage of 18 ppm would deliver approximately 10.6 mg/L dry-basis alum. The calculation is as follows: feeding 18 ppm of alum–polymer blend into the raw water stream is the same as feeding 18 gal of product per 1,000,000 gal of water. An alum–polymer blend weighs about 10.9 lb/gal (it may vary), so 18 gal × 10.9 lb/gal = 196.2 lb/mil gal and 196.2 lb/mil gal/8.34 = 23.5 mg/L of liquid alum–polymer blend. Because roughly 45 percent of this is dry alum, 45 percent × 23.5 = 10.6 mg/L dry-basis alum. Because more dry-basis alum is usually required to remove total organic carbon than is needed to remove turbidity, operators should resist lowering blended dosages.

Table 5-2 shows feed amounts of alum–polymer blends needed, in milliliters per minute and gallons per day, to achieve various parts-per-million dosages.

Iron-based coagulants, such as ferric chloride and ferric sulfate, are also used in water treatment. In some cases, they may be a better coagulant choice because of their ability to work in a wider range of pH. Because iron salts tend to consume more alkalinity than do aluminum salts, they may need less mixing or less contact time for reactions to take place but may depress the pH further. The difficulty that is most often cited when using ferric and ferrous coagulants is their corrosiveness to the storage vessels and feed systems, which impacts the maintenance efforts of operators.

Mixing Facilities—Rapid Mix

Mixing is an important part of the coagulant process, regardless of the coagulant used or the method for feeding the coagulant. Rapid mixing describes the intense mixing that usually takes place in the first moments when the coagulant is fed into the raw water. It also describes the equipment that

Table 5-2 Amount of alum–polymer blend needed to achieve various dosages at several raw water flow rates (mL per min/gal per day)

ppm Desired	20 mgd	40 mgd	60 mgd	80 mgd	100 mgd	120 mgd	140 mgd
15	789/300	1,578/600	2,367/900	3,156/1,200	3,945/1,500	4,734/1,800	5,523/2,100
18	947/360	1,894/720	2,840/1,080	3,787/1,440	4,734/1,800	5,680/2,160	6,628/2,520
21	1,105/420	2,210/840	3,314/1,260	5,302/1,680	5,523/2,100	6,628/2,520	7,732/2,940
24	1,262/480	2,525/960	3,788/1,440	5,050/1,920	6,312/2,400	7,574/2,880	8,837/3,360

is used for the intense mixing period. The purpose of rapid mixing is to briefly and violently agitate the water and coagulant so that particles have numerous opportunities to collide. This process can easily fail in two ways: if the length of agitation is too short or too long or if the mixing energy is insufficient. To overcome this possibility, engineers have devised several facilities, described here, for rapid mixing.

Mechanical mixers—propeller, impeller, or turbine type—can be placed in square or rectangular chambers. The detention time in these chambers is very short—usually 15 to 45 seconds. Most plants are equipped with multiple units so that operators can take them in and out of service. In this way, the range of desired detention times can be achieved. Operators should closely monitor these times and make every effort to maintain them. Units are normally equipped with gates or valves for ease of control. If a mechanical mixer is placed directly into the pipe, it is referred to as an *in-line mixer*. In either case, the chemical application point must be located at or very near the mixer blades to allow for quick dispersion of chemical.

Static mixers provide good mixing by placing fixed vanes inside of the pipes carrying the raw water flow. These units are economical and relatively free of maintenance requirements but do increase the head loss in the system.

Baffled chambers are sometimes used to provide mixing for chemical applications. They are limited in their ability to provide turbulence, which is controlled by the rate of raw water flow.

Pumps used for feeding chemicals sometimes double as mixers. The coagulant is added to the suction side of the pump; the pump itself mixes the chemicals. Like the baffled chamber, there is little control over the amount of mixing.

FLOCCULATION

After coagulation, the newly formed floc particles are brought into the flocculation stage, where gentle mixing allows them to grow in size. This physical process is also governed by detention times and mixing energies. The process of particle size growth in flocculators is sometimes called *agglomeration*. As particles grow in size in the flocculators, they become more fragile. For this reason, flocculators are usually baffled and provided with tapered mixing (Figure 5-6).

The baffling creates compartments in the flocculators that reduce short-circuiting. Most modern flocculators are equipped with two or three sets of under/over baffles, which create three or four compartments. As water flows under and over the baffle walls, mechanical mixing is provided to each compartment. Each successive compartment receives less mixing energy than the previous compartment. This allows the floc particles to grow in size yet receive progressively gentler treatment.

Source: Conneaut, Ohio, Water Department.

Figure 5-6 Flocculator drive—variable-speed unit

This is accomplished in one of two ways. Separate motors and shafts of the same size can be supplied for each compartment and the size of the mixer blades is reduced from one compartment to the next. These units are vertically mounted. Alternatively, one motor and one shaft can be arranged horizontally through the basin's compartments. The shaft has paddles fixed to it; the first compartment may have four paddles, the next three, and the final, two paddles. The shaft turns at the same speed, but less energy is imparted where there are fewer paddles.

Like rapid mixers, flocculators are usually provided in multiple sets, and operators should keep the proper number of units in and/or out of service to provide the correct detention times (Figure 5-7). Flocculators are designed to provide 15 to 45 minutes of detention time. A common operational mistake is to keep too many units in service, thereby providing excessive flocculation time. This can break up fragile flocs, creating a need for excessive amounts of coagulant. It is not uncommon to see flocculators with large amounts of settled floc because the detention time was too long. This also creates a maintenance problem—units have to be cleaned more often. If they are not cleaned, short-circuiting occurs and the problem worsens.

Source: Philadelphia Water Department.

Figure 5-7 Flocculators at Philedelphia Water Treatment Plant out of service for maintenance

In addition to operating the proper number of flocculators, operators must control the mixing energy imparted to the floc chambers. Control is designed into most modern flocculators so that an operator can choose the "tip speed," at least for the final stage. A good rule of thumb is that the tip speed of the final flocculator paddles should be about 1 ft/sec and operate in the range of 0.9 to 1.3 ft/sec most of the time (Figure 5-8). In colder water or when floc formation is heavy due to naturally occurring turbidity, a faster speed may be needed. The tip speed of the final paddle is determined using a stopwatch to measure the time it takes for the paddle to make one revolution. If the distance from the center of the shaft to the end of the paddle is measured, the distance traveled in that one revolution can be calculated as $\pi \times D$, where D is the diameter of the circle, in feet, that the paddle makes when it revolves. D is calculated as 2 times the radius. The distance is divided by the time to obtain the tip speed, in feet per second. For example, if the final paddle in a flocculator revolves once every 55 seconds and the distance from the center of the shaft to the tip of the paddle (radius) is 11 ft, what is the tip speed? Distance traveled by the paddle is $\pi \times D$, or 3.14×22 ft, which is 69 ft; 69 ft divided by 55 seconds is 1.25 ft/sec.

Source: Philadelphia Water Department.

Figure 5-8 Flocculator drive at Philadelphia Water Treatment Plant used to control paddle speed

Combined Units

Some process units combine rapid mixing and flocculation in the same unit. Referred to as *solids contact units* or *contact clarifiers*, these units are generally circular in shape and contain equipment for mixing and sludge withdrawal. There is more discussion on these units in chapter 6.

BIBLIOGRAPHY

AWWA. 2010. Principles and Practices of Water Supply Operations—*Water Treatment*, 3rd ed. Denver, Colo.: American Water Works Association.

AWWA. 2011. M37—*Operational Control of Coagulation and Filtration Processes.* Denver, Colo.: American Water Works Association.

Edzwald, J.K., ed. 2011. *Water Quality and Treatment*, 6th ed. New York: AWWA and McGraw-Hill.

Ross, G., T.F. Zabel, and J.K. Edzwald. 1999. Sedimentation and Flotation. In *Water Quality and Treatment*, 5th ed. Raymond D. Letterman, ed. New York: McGraw-Hill.

Routt, J.C. 2001. Personal communication. October.

Chapter 6

Sedimentation

In chapter 5, mixing and coagulation were identified as processes that are operated as barriers against the passage of particulate contamination. Proper operation of these barriers was explained with regard to design considerations: An operator uses the right amount and types of chemicals for the proper amount of time, and with sufficient mixing energy, and expects good results.

In this chapter, sedimentation as a unit process barrier is examined. Here, too, the process has specific design and operating parameters that, if followed, will help the operator produce optimized results. In addition, *performance goals* related to sedimentation will allow the operator to determine the effectiveness of the operation. Performance goals are water-quality-oriented results that operators expect to achieve most of the time under all operating conditions. These goals have common distinctions. Any operator on shift easily determines these values at any time; they are realistic for the plant operation, i.e., they are reasonably achievable; they are written and published for all to see; and they possess the "buy-in" of the entire staff and administration. In fact, the goal-setting process is a means to an end. Plants that set goals seem to outperform those that do not.

The performance goal normally associated with the sedimentation process is production of a sedimentation-basin effluent with low turbidity. Particle counts are also used to set goals for sedimentation. Because the purpose of a sedimentation basin, or clarifier, is to reduce the incoming turbidity to a fraction of its original value, goals are normally set based on the occurrence and magnitude of raw water turbidity.

It is important to note that to optimize sedimentation, goals need to be established for each sedimentation basin rather than relying on goals for a composite. Often when there are several sedimentation basins, one may have problems but the composite results make detection difficult.

An examination of the performance goals that have been set by the Partnership for Safe Water illustrates this point. The partnership goals for sedimentation are

- Settled water turbidity with less than 1 ntu 95 percent of the time when raw water turbidity is less than or equal to 10 ntu.

- Settled water turbidity less than 2 ntu 95 percent of the time when raw water turbidity is greater than 10 ntu.

Other sedimentation performance goals may be important to operators. It is not uncommon for operators to set sedimentation performance goals (reductions) for iron and manganese, oxidant demand, total organic carbon (TOC), heterotrophic plate count, and color. Softening plants will have performance goals that relate to hardness and alkalinity (see chapter 9).

These goals suggest an important operational philosophy: Regardless of the incoming raw water contaminant, the contaminant output from the sedimentation basin should not deviate significantly and should always be minimal, especially for particulates.

PROCESS DESIGN

Sedimentation (clarification) depends on gravity to remove, or separate, solids from water. Basins constructed for this purpose can be rectangular, square, or circular. In a conventional water treatment plant, sedimentation is situated after flocculation and before filtration. The basins are designed to allow water to flow through them very slowly, and care is taken to ensure minimum turbulence at the inlet and outlet (Figure 6-1).

Various manufacturers' designs are available to minimize short-circuiting through the basin; however, operators should expect these designs to be only partially effective. The sludge, or residuals, that accumulates on the bottom of the basin must be periodically or continuously removed to waste. The outlet of the basin is usually equipped with weirs, or launders, that collect the clear water from the top of the basin and carry it to the filters. A schematic of a typical sedimentation basin is shown in Figure 6-2.

Simple clarifiers have basic design parameters that operators should understand. They are built with a minimum flow-through time or detention

Source: Philadelphia Water Department.

Figure 6-1 Sedimentation basin inlets at Philadelphia Water Treatment Plant. Inlets are designed to reduce velocity and to provide even flow across the basin.

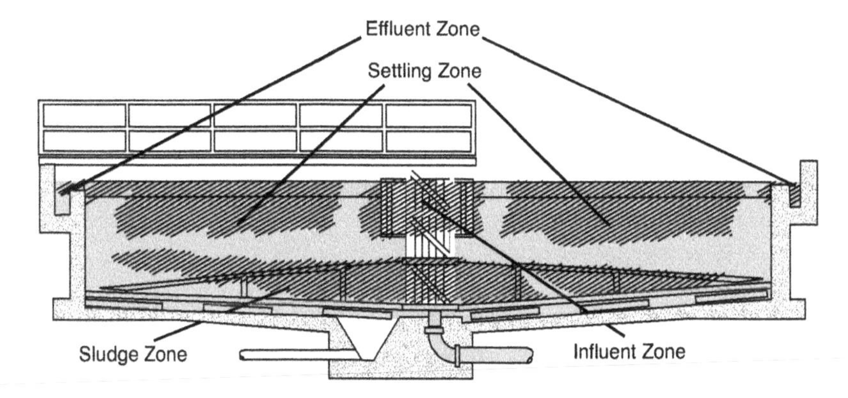

Figure 6-2 Schematic of sedimentation basin zones

time. This detention time is the theoretical amount of time that a quantity of water would take to pass through the basin and exit at the outlet. Most clarifiers are constructed to have 4 hr of detention time at the maximum flow rate of the plant, although groundwater-softening plants may have basins that are designed with less detention time. Detention time is calculated using the following formula:

$$\text{detention time} = V/Q \qquad \textbf{(Eq. 6-1)}$$

where Q is the flow rate, in cubic feet per minute, and V is the volume of the tank, in cubic feet.

To learn theoretical detention time calculations, operators can envision the amount of time that it takes to fill or empty a tank at a given flow rate. For example, at a flow of 5 mgd (7.75 ft^3/sec), how much time does it take to fill a rectangular tank to a depth of 20 ft (6 m) if that tank is 100 ft (30.5 m) long and 58 ft (17.7 m) wide, starting from empty? Using Eq. 6-1, convert the flow rate from cubic feet per second to cubic feet per minute. This is done by multiplying 7.75 ft^3/sec (3,480 gpm) by 60, which yields a flow rate of 465 ft^3/min (15.3 L/sec). The volume of the tank is computed by multiplying the length, width, and depth, which yields a volume, in cubic feet, of 116,000. Finally, divide the volume by the flow (116,000/465) to obtain a detention time of about 249 min, which is just over 4 hr.

The above example calculation can be misleading. As discussed in the previous chapter, tanks such as clarifiers and sedimentation basins exhibit considerable short-circuiting, i.e., some of the water entering the tank escapes to the other end in a time much shorter than the theoretical detention time. Detention time is an *average* time for the water to flow through the tank. Operators need to understand that changes made to the water entering the tank in the above example will not necessarily take 4 hr to exit the other end. This is important to understand. A mistake made in coagulation, for example, is likely to impact the filters in a time much shorter than 4 hr.

Another useful design parameter for clarifiers is the surface overflow rate (SOR), which is given in gallons per minute per square foot (meter per hour). Simple basins are usually designed for an SOR of 0.5 gal/min/ft^2 (1.2 m/hr). Basins that are operated in excess of the design SOR tend to load the filters with greater amounts of suspended solids, which can reduce filter run times and increase the probability of turbidity breakthrough in filters. The formula used to calculate the SOR is

$$\text{SOR} = Q/\text{square feet of surface} \qquad \textbf{(Eq. 6-2)}$$

where Q is the flow rate, in gallons per minutes, and square feet is the area of the basin. Note that the depth of water in the clarifier is not used in this calculation. This design parameter works for circular as well as rectangular basins. In the previous example, the SOR would be calculated by multiplying 100 ft by 58 ft (30.5 m by 17.7 m), which is 5,800 ft^2. The flow rate, Q, is 5 mgd, or 3,472 gpm (464 ft^3/min). The SOR then is 0.6 gpm/ft^2 (3,472/5,800) (1.47 m/hr). Properly designed sedimentation basin launderers help to control SORs (Figure 6-3).

In water treatment plants with multiple basins, operators should carefully manage the process so that these two design parameters are not exceeded. All sedimentation basins must be taken out of service periodically for cleaning and maintenance, even if the units are provided with automatic sludge-removal equipment. When basins are removed from service, operators should calculate the resulting detention times and overflow rates of the remaining basins and adjust the operation as necessary. A basin outage should be a planned event, and all operations staff should be aware of its ramifications.

TYPES OF BASINS

Conventional sedimentation basins have a rectangular-feed or center-feed configuration. Rectangular basins may be constructed of concrete or steel and are designed so that flow velocity is parallel to the basin length. The velocity through the basin should not exceed 0.5 ft/min (0.15 m/min). To achieve even flow distribution across the basin, inlets are provided that, when kept clean and free of impediments, will minimize the eddies and currents that cause short-circuiting. Outlet weirs that can also help to minimize short-circuiting are provided. The purpose of the weir is to collect clear water from the top portions of the basin for transport to the filters. Weir overflow rates (not to be confused with SOR) are designed at about 20,000 gpd/ft (250 m^3/day/min). An overflow pipe is provided so that a maximum water level is maintained and not exceeded; this provides a maximum water level to the filters. Basins will have various sludge collection and removal systems (see discussion later in the chapter).

Conventional center-feed basins can be round or square. The settled water flows from the center to the outside (radial flow) and, like rectangular basins, these basins are designed to keep an even flow distribution. Basin

Source: Philadelphia Water Department.

Figure 6-3 Sedimentation basin launderers at Philadelphia Water Treatment Plant arranged to reduce solids carryover onto filters

bottoms are conical, which helps the sludge-collection process. Center-feed basins can be peripheral- or spiral-flow.

Sedimentation basins have four zones, each with a distinct function. The *influent zone* decreases the velocity of the incoming water and distributes it evenly. The *settling zone* provides a calm area, which is necessary for suspended particles to settle. The *effluent zone* is the area where the transition from settling to effluent flow takes place. Eddies and currents that could stir up sediment from the bottom of the basin must be minimized in this zone. Finally, the *sludge zone* is the area where sludge particles are accepted and kept separate from other particles in the settling zone. These zones are not well defined, i.e., it is difficult to determine where one zone stops and the other starts. The zones are fluid and can shift from area to area. It is important to know that the settling zone is always affected by the actions in the other three zones and the quality of the water leaving the basin will be determined by those actions.

A clarification unit that combines coagulation, flocculation, and sedimentation in a single basin is called a *solids-contact unit* (Figure 6-4). Typically, these units rely on the recirculation of sludge into the process and

Source: US Filter.

Figure 6-4 Solids-contact unit with center feed and mixers

are usually operated to maintain a sludge blanket. These units are used in the lime-soda process because the chemistry of softening is enhanced by contact with previously precipitated calcium carbonate. A solids contact unit must be properly operated to maintain a good sludge blanket. This is accomplished by systematically withdrawing a portion of the sludge. Automatic timer controls are usually provided, and the operator must provide the necessary settings. These units, designed to operate at SORs of 1.0 gpm/ft^2 (2.4 m/hr), may be designed for even higher rates if the unit is used for lime softening. The main drawbacks of these units are sensitivity to increases in flow rates and to temperature variations. Both conditions can upset the sludge blanket and result in floc carryover.

HIGH-RATE PROCESSES

Several variations of the clarification process have been developed to allow the operator to produce water at higher rates than are possible with conventional processes. Units that can operate at higher SORs accomplish the task in a satisfactory fashion and may take up less space in the plant. Most processes provide shorter response times for operators to correct treatment problems, but overall maintenance efforts and demands are comparable to conventional processes.

Tube and Plate Settlers

Particles suspended in water must settle to the bottom of a basin in order for the sedimentation process to be effective. In conventional settling basins, this often means that a particle must travel 15 ft (4.5 m) or more before it is truly removed from the process. As gravity pulls the particle to the bottom of the basin, it resists settling because of the horizontal flow of water, and so may be carried to the filters. Consequently, basins must be large enough to provide long detention times.

One way to reduce the distance that a particle must travel is to insert tubes or plates into the basin at angles; a practice that increases the effective surface area of the unit. The result is shallow-depth sedimentation, which can increase the effective SOR of the basin two to four times. Typical SOR rates for basins equipped with tubes or plates are 2.0 gpm/ft^2 (4.8 m/hr). Tubes or plates are designed as either horizontal or as steeply inclined, depending on the application (Figure 6-5). Horizontal tube settlers are actually slightly inclined (perhaps 5°) and work dependently with the downstream filter. Sludge accumulates in the tubes until the filter is backwashed. When the filter is drained, the water level in the settler is lowered and the sludge is carried to waste. This application is frequently used in small package plants.

Source: US Filter.

Figure 6-5 Inclined plate separator

Steeply inclined tubes or plates are designed into basins with varying degrees of incline. When they are inclined at less than 45°, the method of sedimentation is generally capture and storage. For this reason, these units must be cleaned periodically to prevent over accumulation of solids. In some plants, tube settlers have accumulated enough solids to collapse under the weight.

Cleaning is usually accomplished by slowly lowering the water level in the basin, allowing the solids to slough to the bottom where they are wasted. At an incline greater than 45°, solids do not accumulate on the tubes but rather move down the tubes in a continuous fashion. Tubes and plates are built and shipped as modules of varying size and therefore lend themselves to retrofitting into existing basins. To gain more capacity, many water treatment plants have resorted to this type of retrofit rather than building new and expensive sedimentation basins.

These arrangements rely on an upflow direction of water, i.e., the water to be settled is introduced to the bottom of the plates or settlers and allowed to flow up through them. If operated properly, the water velocity is low enough to allow the solid particles to settle. Another arrangement is to install parallel plates into the basin with the water to be settled directed downward to the bottom of the basin and then upward into the plates. These plates, called *Lamella plates*, are usually inclined at 55°.

Superpulsators

A superpulsator is a modification of a sludge blanket clarifier coupled with Lamella plates, which enhances coagulation. SORs as high as 5 gpm/ft^2 (12 m/hr) have been reported for these units. The water entering the clarifier is "pulsed" by the application of a vacuum created in an upstream vacuum chamber. The vacuum pulls the water into the chamber and, when the water rises to a certain level, the vacuum is released. The water then surges into the clarifier through the distribution conduit and laterals, causing the sludge blanket to expand uniformly. The chemically treated water must pass through the sludge blanket, where particle contact takes place. With each surge, some of the residuals are allowed to overflow into a sludge concentrator zone where they are thickened and sent to waste. The supernatant flows up through the Lamella plates, which have been modified with deflectors. The surges create small eddy currents that are redirected by the deflectors into the flocculated sludge. The heavier particles in this zone settle out, and the clarified water is collected and sent to filtration.

Actiflo® Process

The Actiflo® (US Filter, Warrendale, Pa.) process has separate mixing, coagulation, and sedimentation compartments (Figure 6-6). The process uses microsand, which enhances flocculation and settling. Floc particles adhere to the microsand and are removed from the center hopper in the sedimentation

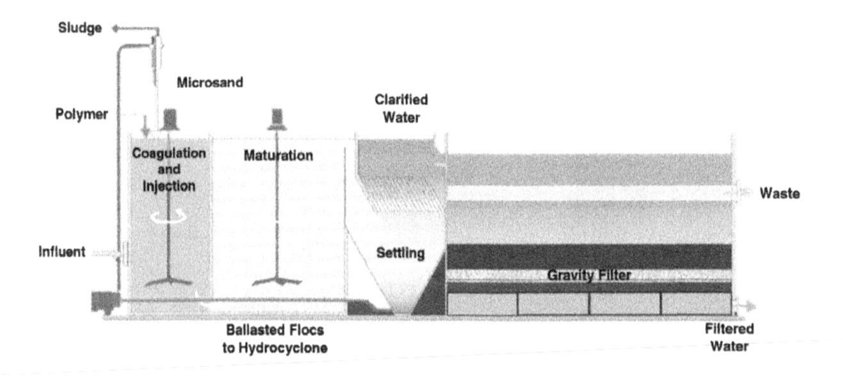

Source: US Filter.

Figure 6-6 Schematic of a preengineered water treatment plant

compartment. These solids are pumped to a hydrocyclone compartment where solids are separated from the microsand and carried to waste. The microsand then sinks to the bottom and is reused. When operated properly, these units can produce low-turbidity effluent at SORs up to 16 gpm/ft^2 (39 m/hr).

Dissolved Air Flotation

Dissolved air flotation (DAF) is a high-rate process that has been successfully applied in many parts of the world for decades. Like most high-rate processes, it has more system components compared to a conventional facility; however, it is reliable and robust and does not necessarily require more operator sophistication. In fact, numerous utilities have recently installed DAF systems to replace conventional plants and have made the transition rapidly and more simply than anticipated. DAF systems typically operate at hydraulic overflow rates in the range of 6 to 10 gpm/ft^2 (14.7 to 24.5 m/hr).

DAF involves rapid mixing and flocculation followed by clarification; sludge is brought to the top of the basin, rather than to the bottom, as with conventional treatment (Figure 6-7a). This is accomplished by recycling a portion of the clarified water (about 10 percent), saturating it with air under pressure (roughly 70 psi), and delivering it to the base of the flotation tank where a pressure drop occurs through a series of nozzles or valves. Due to the pressure drop, air is released from solution and mixed with the incoming flocculated water (Figure 6-7b). The bubbles attach to the floc, which then rise to the top and rear of the tank over an inclined baffle. A sludge, or "float," layer develops at the top of the flotation tank, which is periodically removed by mechanical (skimmer) or hydraulic means. The objective of DAF is to produce small, pin-sized floc that is readily amenable to flotation.

Source: ITT Water and Wastewater Leopold Inc.

Figure 6-7a DAF unit schematic

DAF was originally used to treat cold low-alkalinity waters that were prone to algal blooms. DAF has also been successfully used to treat water with more diverse qualities, especially recycled backwash water, which tends to be highly turbid and variable.

In a pilot-scale study performed at the Cleveland, Ohio, Division of Water, Morgan Water Treatment Plant, DAF was shown to treat spent filter backwash water (SFBW) at a hydraulic overflow rate of 6 gpm/ft^2 (14.7 m/hr). This was eight times the rate employed by conventional clarification with polymer addition in the parallel study and comparable to rates used in potable water treatment. The average untreated turbidity of the SFBW was 25 ntu and ranged between 10 and 60 ntu during the study. In the study, conventional clarification with polymer (at an overflow rate of 0.8 gpm/ft^2 [2 m/hr]) reduced the clarified effluent turbidity to 2 ntu, while the DAF system reduced it to less than 1 ntu on average.

Due to its ability to remove low-density particles, DAF has achieved a 2-log removal of *Cryptosporidium* oocysts, which compares to a removal rate of less than 0.5 log using conventional sedimentation.

Source: ITT Water and Wastewater Leopold Inc.

Figure 6-7b DAF overview with air tanks

OPERATIONS AND MAINTENANCE

Sedimentation is an important barrier against particulate passage into the filters. When designed and operated properly, clarifiers remove a tremendous amount of particulates in a relatively short period of time. It is not uncommon for sedimentation basins to remove 90 percent or more of the source water turbidity. Conventional basins tend to be very expensive to design and construct. As a result, in larger water treatment plants, conventional basins barely meet minimum design standards for detention time. When taken out of service, they represent a large percentage of capacity loss. In a small plant with only two sedimentation basins, 50 percent of the capacity is lost each time a basin is down for service. Explained another way, 50 percent of the process that removes 90 percent of the solids is temporarily lost in that instance.

As previously mentioned, sedimentation basins should be cleaned manually, preferably twice each year. Most plants schedule this maintenance just before and just after the summer high-pumping months. Plant operators in northern climates prefer not to remove basins in winter because colder water makes settling more difficult (Figure 6-8). Also, the chilled basin walls make it very uncomfortable for operators to work, even when the basins are covered.

Manual cleaning is important because harmful and nuisance bacteria and organisms can collect in the slimes that adhere to the basin walls and appurtenances. These slimes, or biofilms, accumulate on rough surfaces and in crevices where chlorine has difficulty reaching. Particularly vulnerable are the interfaces between the water surface and the air surrounding it. Ever-present scum lines, or "rings," make an excellent shield for organisms such as Legionnaires' bacteria. Operators need to blast these areas with water from high-pressure hoses to release the bacteria. Operators must wear paper masks to prevent inhalation of the mists generated during blasting.

To reduce chlorinated by-products, many plant operators have reduced or eliminated prechlorination. At these plants, it may be wise to periodically chlorinate the basin for a few hours. This maintenance procedure, referred to as "shock" by operators, is typically performed in uncovered basins that are exposed to direct sunlight. These basins tend to accumulate algae that can cause taste-and-odor problems. Another way to rid a basin of slime growth is to coat the walls with a mixture of copper sulfate and lime, usually at a concentration of 10 g/L. The basin is drained and the mixture applied to the walls with a brush.

Figure 6-8 Floc carryover and ice formation on solids-contact unit. Heavy floc carryover can reduce filter run length.

Sludge often accumulates on the basin floor, even when automatic sludge-removal devices are used. Most basins have areas where cleaning equipment cannot reach. Sludge can age in these spots and "go septic," i.e., become devoid of oxygen and produce foul-smelling gases. This is to be avoided.

It is a good idea to perform physical inspection of the basins while they are emptied. Basins with notched launderers should be checked to see that the troughs have not become misaligned due to freezing conditions. Unleveled troughs will create hot spots of localized high overflow rates and draw too many solids from the depths of the basin. This can cause filtration problems. If catwalks or walkways are a part of the basin, operators should check for rust or paint chips or grease that can drip and contaminate the water. This is also a good time to maintain any moving parts.

SAFETY

Safety measures are important when cleaning sedimentation basins. Most interpretations of Occupational Safety and Health Administration rules classify sedimentation basins as confined spaces. Consequently, these rules would apply to cleaning procedures of or for any entry into the unit. The use of high-pressure water hoses can present a hazard to the operator. Wet, slippery floors and the loss of control of a hose are dangerous to the operator and any other occupants. Also, because unexpected flow of water into the basins from the inlet source is a form of energy release, many operators follow lockout–tagout procedures before entry. An operator working in a deep basin with his or her back to the wall may not be able to hear the rush of water that results from an inadvertent valve opening until it is too late. Many plants keep a life ring or pole handy in case a worker falls into the basin. Moving parts, especially in a chain-and-flight system, present a hazard to workers. Such machinery should have guards placed on them. Guardrails that surround the basins are required. Safety practices are explained further in chapter 14.

RESIDUALS

All basins, whether conventional or high-rate, require sludge disposal. This is usually the most troublesome operation in a water treatment plant. In the past, operators commonly discharged residuals to a stream or lake. However, discharge regulations, both for the environment and for wastewater plants, now restrict an operator's ability to waste residuals with impunity. The process of sludge disposal is now as tightly controlled as any other process in the plant. Residuals are removed manually or with an automatically controlled vacuum or raking systems (Figure 6-9).

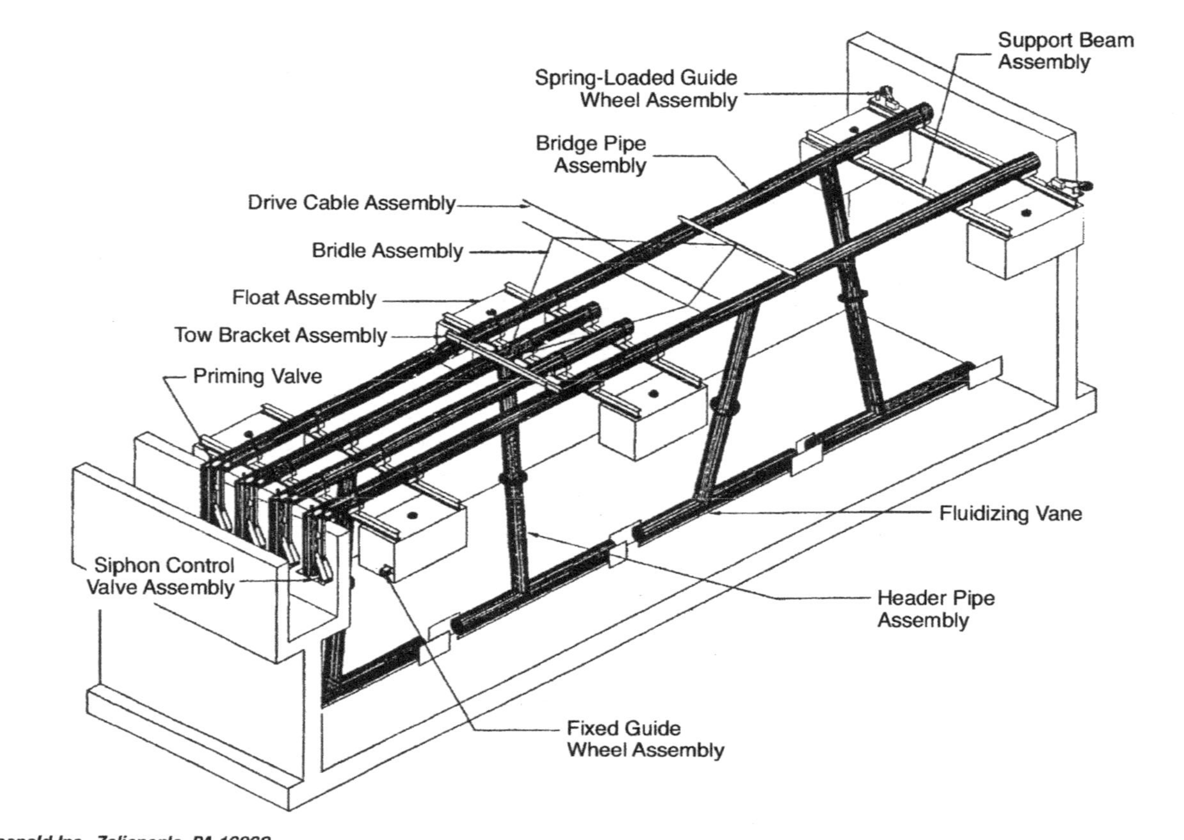

Source: F.B. Leopold Inc., Zelienople, PA 16063

Figure 6-9 Floating sludge collector

Residuals are comprised of all the unwanted material that has been removed from the water and all of the compounds that are created by addition of treatment chemicals (floc, lime solids, etc.) (Figure 6-10). Because this chapter is about sedimentation, residuals from brines used in ion exchange or wastes from membrane applications are discussed in other chapters.

Alum sludge is the most common residual produced in water plant sedimentation basins, but ferric sludge also is common. Both wastes come from the sedimentation basins at low concentrations by weight: normally less than 2 percent. It is inefficient to dispose of this low concentration of sludge, and so water plant operators dewater and concentrate the residual.

Dewatering is the process of removing water from the thin sludge, allowing the solid material to remain (thickening) for ultimate disposal. It is difficult to dewater alum and ferric sludge because the water is chemically bound to the aluminum and ferric hydroxide floc. These flocs are gelatinous and thus difficult to handle. As a rule of thumb, the thicker the sludge, the less space it takes up, and so less storage space will be needed. Additionally, the thicker the sludge, the less expensive it is to dispose. Wastewater plants and landfills usually charge by volume of material they receive (there may be other surcharges for strength of solids or for special considerations due to the nature of the sludge). When disposing of thin sludge, therefore, the water plant operator is also paying to dispose of a larger proportion of water than may be necessary.

Sludge drawn from sedimentation basins varies with the coagulant used, of course, but also with the nature of the material found in the raw water. Sludge produced in plants that draw water from lakes and other sources with relatively low turbidity is very gelatinous, whereas sludge produced from a source high in silt and organic content is not as gelatinous and so more easily handled. The percent solids concentration of the sludge also affects its characteristics—low-solids sludge is more liquid than high-solids sludge, which tends to handle like clay.

In many water treatment plants, sludge from sedimentation basins is pumped to a gravity thickener that is designed to dewater the waste to a specific concentration before disposal. This gravity system allows the solids to settle to the bottom and provides for removal of the clear supernatant at the top of the unit. Supernatant is either returned to the process or sent to waste. The thickener also serves as an equalization tank, allowing for continuous or intermittent wasting of solids that might otherwise burden the receiving sewers or treatment facility.

Further treatment of sludge after thickening can take place in the plant or at the disposal site (Figure 6-11). Water treatment plants often treat thickened sludge with polymers or lime and may subject the sludge to flotation or belt-press thickening to obtain the desired concentration. When sand drying beds, lagoons, or freeze–thawing beds are employed, the process is referred to as *nonmechanical dewatering*. Mechanical dewatering is usually

Source: Philadelphia Water Department.

Figure 6-10 Sludge collectors at Philadelphia Water Treatment Plant. Rakes help to concentrate the solids.

accomplished with vacuum filtration, belt presses, or centrifuges. All of these processes are effective but costly. Each water treatment plant needs to determine the most cost-effective procedure for its situation, taking into account all economic factors at its disposal. Local regulations will also be a factor.

For simplicity and operator convenience, the sand-drying bed method of solids handling is difficult to beat. Costs of the process vary by site and size, but beds with greenhouses built around them to trap heat, and with fans to hasten drying, show great promise for freeing plant operations from the burgeoning costs imposed on them for residuals disposal when discharging to wastewater systems.

Some utilities with warm and sunny climates have been using simple sand beds for years. Ongoing research is leading engineers to design these systems for utilities in varying climates. One process refinement is simple in its application: thickened sludge is pumped to the bed as polymer is injected into it. Sludge is loaded onto the beds at rates of about 3 dry lb of solids per square foot of bed. Polymer dosages are typically found to be 3 lb of polymer per dry ton of sludge. The cakes that are formed can quickly dry to 30 percent or more in a few days, and are easily handled and disposed.

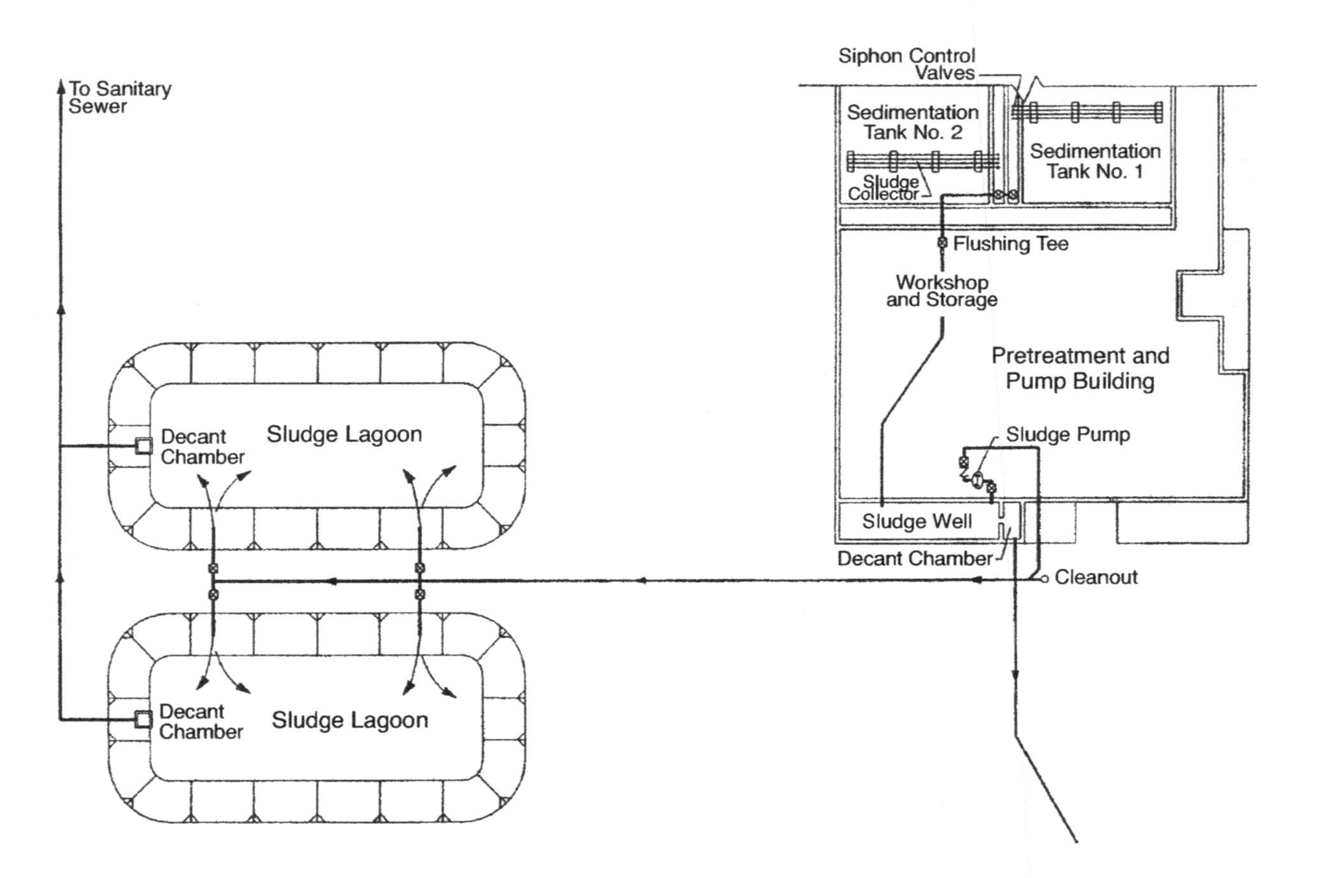

Figure 6-11 Typical sludge lagoon system. Lagoons can be operated independent of each other, allowing for maintenance.

Sludge Calculations

If the concentration of solids coming from the sedimentation basins is known, the amount of solids that the thickener will produce can be calculated using the formula:

$$V_2/V_1 = P_1/P_2 \qquad \textbf{(Eq. 6-3)}$$

where V_1 and V_2 are the beginning and ending volumes of the sludge and P_1 and P_2 are the beginning and ending concentrations of sludge, expressed as percent. This formula assumes that the specific gravity of the sludge varies little in typical dewatering processes. As an example, suppose that 100,000 gal of sludge is removed from a sedimentation basin at 0.2 percent (2,000 mg/L) solids and is sent to a thickener where it will be concentrated to 5 percent (50,000 mg/L) before being transferred to a tank truck for hauling. If the tank truck has a 4,000-gal capacity, how many trips will be necessary?

Answer: $V_2 = P_1 \times V_1/P_2 = 0.2 \times 100{,}000 \text{ gal}/5 = 4{,}000 \text{ gal}$

One truckload will be sufficient.

Operators use other formulas to predict the amount of sludge that will be produced by the chemical coagulation and/or precipitation processes. These formulas are for alum and ferric salts and for softening chemicals such as lime and soda ash.

$$\text{For alum, } S = (8.34Q)(0.44\text{Al} + SS + A) \qquad \textbf{(Eq. 6-4)}$$

where:

S = sludge produced, in pounds per day
Q = plant flow, in million gallons per day
Al = dry alum dose, in milligrams per liter as 17.1 percent Al_2O_3
SS = raw water suspended solids, in milligrams per liter
A = solids from other treatment chemicals, such as powdered activated carbon, etc., in milligrams per liter

$$\text{For iron salts, } S = (8.34)(2.9\text{Fe} + SS + A) \qquad \textbf{(Eq. 6-5)}$$

where Fe is the iron dose, in milligrams per liter as Fe.

$$\text{And for softening, } S = (8.34\ Q)(2.0\text{Ca} + 2.6\ \text{Mg}) \qquad \textbf{(Eq. 6-6)}$$

where Ca and Mg are the calcium and magnesium hardness removed, as milligrams per liter calcium carbonate.

Example calculation:

The *SS* portion (raw water suspended solids) of the preceding formulas is either measured in the laboratory or approximated using a "*B* value." The *B* value is a ratio of the suspended solids in the raw water to the turbidity of the raw water. It is usually found to be about 0.7 to 2.2 but can be much higher if large amounts of TOC are present. An operator's ability to quickly analyze for turbidity makes the *B* value a great help in predicting solids production. Each water treatment plant needs to develop a history of solids measurement in order to rely on a usable *B* value.

Using Eq. 6-4, calculate the pounds of solids produced per day if a treatment plant processes 25 mil gal of water using 15-mg/L dry-basis alum. The raw water turbidity is 44 ntu and the historic *B* value for this plant is 1.5. The plant is adding 5 mg/L activated carbon for taste-and-odor control.

Multiply the *B* value by the *SS*: $1.5 \times 44 = 66$. The calculation becomes

$$(8.34 \times 25)((0.44 \times 15) + 66 + 5) = (208.5) \times (77.6) = 16{,}180 \text{ lb}$$

TURBIDITY TESTING

As previously mentioned, the performance of a sedimentation basin is commonly measured using turbidity output. It is a good practice to determine the turbidity from each basin, rather than from the combined output of a series or bank of basins. A poorly performing unit can be masked by the dilution of good water from the other basins in the bank. Some operators prefer to measure the incoming and outgoing turbidity or particle count of each basin. The difference in the two readings is measured against a preset goal, either as a percent removal or as a log removal. When the goals are not met, remedial action is indicated.

BIBLIOGRAPHY

AWWA. 2010. Principles and Practices of Water Supply Operations—*Water Treatment*, 4th ed. Denver, Colo.: American Water Works Association.

Cornwell, D.A., M.M. Bishop, R.G. Gould, and C. Vandermeyden. 1987. *Handbook of Practice: Water Treatment Plant Waste Management*. Denver, Colo.: American Water Works Association and AwwaRF.

Cornwell, D.A., M.J. MacPhee, N.E. McTigue, H. Arora, M. LeChevallier, and J.S. Taylor. 2001. *Treatment Options for Removal of* Giardia, Cryptosporidium, *and Other Contaminants in Recycled Spent Filter Backwash Water*. Denver, Colo.: American Water Works Association and AwwaRF.

Cornwell, D.A., and D.K. Roth. 2008. *Altering Environmental Conditions to Enhance Non-Mechanical Dewatering of Water Treatment Plant Residuals*. Denver, Colo.: Water Research Foundation.

EE&T Inc. 2000. Preliminary Evaluation of Morgan WTP Residuals Handling Facilities. Newport News, Va.: Environmental Engineering & Technology Inc. September.

Edzwald, J.K., ed. 2011. *Water Quality and Treatment*, 6th ed. New York: American Water Works Association and McGraw-Hill.

Edzwald, J.K., J.E. Tobiason, H. Dunn, G. Kaminski, and P. Galant, 2001. Removal and fate of *Cryptosporidium* in dissolved air drinking water treatment plants, Water Science and Technology, 43(8):51–7.

HDR Engineering Inc. 2001. *Handbook of Public Water Systems*, 2nd ed. New York: John Wiley and Sons.

MacPhee, M.J. 2001. EE&T, Internal memo to author. October.

MacPhee, M.J., and D. Cornwell. 2000. Critical Assessment of Alternatives for Treatment of Spent Filter Backwash Water. In *Proc. Water Quality Technology Conference*. Denver, Colo.: American Water Works Association.

Partnership for Safe Water, Phase IV Application Package. 2001. American Water Works Association, Denver, Colo. http://www.awwa.org/Portals/0/files/resources/water%20utility%20management/partnership%20safe%20water/files/P4_Revised0311.pdf

Pizzi, N.G., and M.L. Rodgers. 1998. Using Particle Count Data to Improve Operations. In *Proc. Water Quality Technology Conference*. Denver, Colo.: American Water Works Association.

Pizzi, N.G. 2010. *Water Treatment Plant Residuals (Pocket Field Guide)*. Denver, Colo.: American Water Works Association.

Recommended Standards for Water Works. 2007. Albany, N.Y.: Health Education Services.

USEPA. 2002. The Filter Backwash Recycling Rule—Technical Guidance Manual, Office of Ground Water and Drinking Water (4606M), EPA 816-R-02-014 www.epa.gov/safewater,December 2002.

Chapter 7

Filtration

Filtration is the fourth unit process that makes up conventional treatment for particle removal (Figures 7-1 and 7-2); the others include coagulation, flocculation, and sedimentation. Filtration is used in water treatment plants to remove particulate material from the water and store it for eventual disposal. Particulates include those already present in the source water, such as clay and silt particles, bacteria, protozoans, viruses, and organic substances, as well as those generated during the treatment process, e.g., when alum, ferric salts, calcium carbonate, and magnesium hydroxide solids are used in the softening process. In some plants, appreciable amounts of iron and manganese may also be present.

Different types of filtration processes are found in water treatment plants, most commonly slow sand filtration and rapid sand filtration, which are granular bed processes. Diatomaceous earth beds, which use a permeable membrane, are also used.

SLOW SAND FILTERS

A slow sand filter, which is always open to the atmosphere, operates at very low filtration rates and does not accept water that has been treated with a coagulant. These filters, which use small-sized sand, compared to rapid sand units, operate at a low filtration rate. Gravity allows the water to flow downward through the bed. This results in solids being removed and stored almost entirely in the top few inches of the sand bed. This layer of sand becomes laden with particles and dead and living microorganisms and serves as the dominant filter medium as the filter cycle progresses. Filter run lengths for these types of filters are usually 1 to 6 months. When head loss becomes excessive, the filter is cleaned by draining the bed below the sand surface and removing the top inch or two of sand.

Slow sand filters are often considered because of their simple technology—no knowledge of coagulation chemistry is required. Because slow sand filtration is partially a biological process, some operators report conflicting results when prechlorination is used. Many studies have been performed to learn more about this aspect of filtration, but it is left to the

Source: Cleveland, Ohio, Division of Water.

Figure 7-1 Filter gallery at Cleveland's Baldwin Water Treatment Plant

Source: US Filter.

Figure 7-2 Central gravity filter

operator to determine what method is best at that plant. Slow sand filtration is most widely used by municipal water treatment plants (i.e., not plants that serve industry, etc.) where source water quality is stable, notably in upper New York and the New England states as well as in the Pacific Northwest.

A slow sand filter is capable of producing a low-turbidity effluent and can effectively remove or control a wide variety of microorganisms. Studies have shown that slow sand filters removed 85 percent of coliforms and 98 percent of *Giardia* cysts in a source water at startup and performed even better as the bed matured. Although there are exceptions, slow sand filters usually produce effluent turbidities that are less than 0.5 nephelometric turbidity units (ntu) as long as source water remains stable and of good quality. Experts expect an increase in use of these units in the future, especially by small systems.

Slow sand filter design usually provides for filtration rates of 0.016 to 0.16 gpm/ft^2 (0.04 to 0.40 m/hr) when sand sizes of 0.15 to 0.40 mm are specified. Surveys of many US treatment plants reveal that operating head losses fall within the range of 2.5 to 14 ft (0.76 to 4.3 m).

Operations and maintenance for these units include checking and recording head loss, adjusting filter production to provide for system demand, and measuring filtered water turbidities and disinfectant residuals. Cleaning, or scraping, the bed is labor intensive, and several mechanical devices have been constructed to assist the operator in this effort. The sand that is removed is usually cleaned and stockpiled for repeated use. The bed can be scraped for several filter cycles, but eventually the bed must be replenished. This operation is called *resanding*.

RAPID SAND FILTERS

Proper operation of rapid sand filters, currently known as rapid granular-bed filters, usually depends on the passage of pretreated water (see chapter 5) through the bed at rates sufficient to drive the particulate material into the deeper layers. Sand sizes are larger than those of slow sand filters. This arrangement, coupled with the higher filtration rates, generates filter runs that are usually hours to a few days in length, as opposed to months (Figure 7-3).

Unlike the slow sand filter, which uses only the top few inches of the bed as the filtering medium, the rapid sand filter may use the entire bed depth for particulate capture. Consequently, it cannot be cleaned by scraping. Cleaning, or backwashing, is accomplished by introducing water and/or air into the bottom or outlet end of the bed in sufficient quantities and with sufficient flow to fluidize and expand the media so that the stored particulate materials are dislodged. These particulates are then carried away in the stream for disposal or reuse.

Rapid sand filters that are open to the atmosphere are called *gravity filters*. Those that are closed and under pressure are referred to as *pressure filters*.

Gravity Rapid Sand Filters

Typical gravity filters are designed with monomedia, dual-media, tri-media (mixed)-media, or granular activated carbon (GAC) beds. Fine-sand monomedia beds, which were the first rapid sand design used in water treatment plants, were eventually replaced in many installations with the dual-media design. The dual- and tri-media designs allow for greater bed penetration by particulate material and therefore for longer filter runs at higher filtration rates. Deep-bed monomedia filters, with coarse media (1-mm effective size

Source: US Filter.

Figure 7-3 Steel gravity filter

or greater), are also being used where regulatory agencies have approved the design. Local regulatory approval of all media designs is usually required before the design is put into use. GAC beds are used alone or as caps on dual-media filters to remove tastes and odors and to improve organic carbon removal. They may also be used as biological filters. The use of GAC beds is discussed in more detail in chapter 10.

Another attribute of the gravity filter design is that an operator can view the filter bed and backwash actions at any time, which is an enormous advantage over pressure filters. Most operators load gravity filters at rates from 2 to 6 gpm/ft^2 (5 to 15 m/hr) and achieve run lengths of 12 to 80 hours. It is common for these filters to produce effluent turbidities of <0.1 ntu for most of the run. Operators report head losses of 2 to 8 ft (0.6 to 2.5 m) and unit filter run volumes between 5,000 and 10,000 gal/ft^2 (205,000 and 410,000 L/m^2).

Pressure Rapid Sand Filters

Pressure filters are closed tanks or cylinders, i.e., not open to the atmosphere, that contain granular media for filtration (Figure 7-4). This allows the operator to rely on single pumping through the filter and into the system without loss of pressure, as in a gravity filter. This can result in savings in electrical costs. Also, development of negative head and the problems that accompany this phenomenon are avoided with this design. Plants with pressure filters may be easier to automate.

However, there are disadvantages of this design, including the fact that the operator cannot see the top of the bed during filtration or backwash. Also, because the tanks are cylindrical, the width of the filter medium is not uniform, which can lead to uneven backwash patterns.

The operation of pressure granular filters is similar to that for their gravity filter counterparts: water must be pretreated, media size and depth are similar, and head loss developments are comparable. Pressure filters, which

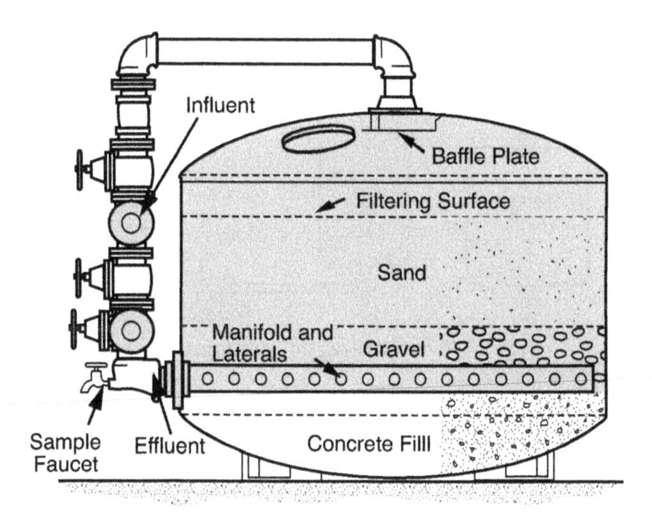

Figure 7-4 Schematic of pressure gravity filter—vertical type

find their greatest use in industrial applications and for swimming pools, are also used sparingly for potable water production, particularly in small systems.

DIATOMACEOUS EARTH FILTERS

Diatomaceous earth (DE) filters (Figure 7-5), also known as precoat filters, are an acceptable filtration technique for potable water production. DE is the skeletal remains of microscopic diatoms or algae. This material is mostly silica and inert, therefore it is suitable to filter water. Because coagulation is not required for operation of these units, operators who possess mechanical skills are more important to the process than those who possess scientific skills. For small systems that must share staff to perform several differing tasks, this can be important.

A high efficiency for removing *Giardia* and *Cryptosporidium* has been reported at facilities that use precoat filters. Other advantages include capital cost savings (precoat filters require only a small space for installation), ease in handling waste residual, and lack of terminal turbidity breakthrough (due to surface, rather than depth, filtration). Disadvantages are that source water quality must be good and the technology may not be applicable for removing total organic carbon.

Process Operation

The process uses DE to strain particulates from water and normally does not use coagulant chemicals. A thin layer of DE coats the filter leaves, or septum, which is a plastic or wire mesh covering a hollow collection channel. The DE coating (precoat) is formed by a recirculating slurry of DE.

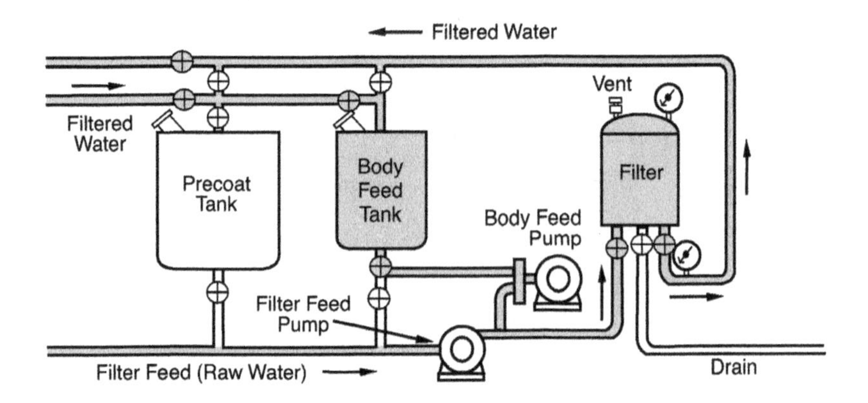

Figure 7-5 Schematic of a diatomaceous filter

There are two main types of DE filter systems: pressure and vacuum. Both have advantages and disadvantages.

Untreated water is applied to the precoat filter (usually about ⅛ in. thick) with a small amount of DE added (called *body feed*). Particulates in the water are removed on the filter surface and this causes pressure to build (head loss). When the terminal head loss is reached, the filter is removed from service for cleaning. High pressure water spray on the filter cake is used to remove the accumulated material. This waste material is discarded and the process is repeated starting with the filter precoat with fresh DE.

DE filtration is used for raw water with a maximum turbidity of 10 ntu. The filter loading rate can be between 0.5 and 2.0 gpm/ft^2. DE filtration is very effective for removing *Giardia* and *Cryptosporidium*. Dissolved substances and color are not effectively removed by DE filtration.

GRANULAR BED DESIGNS

The most common granular filters (see Table 7-1) are monomedia, mixed-media, or dual-media beds that have sufficient depth to produce long filter runs. In the dual and mixed designs, anthracite or activated carbon is used along with sand that has a carefully chosen specific gravity (lower than that of regular sand). This allows the larger-sized anthracite or carbon grains to settle on top of the smaller-sized sand. This large-to-small arrangement provides the longer filter runs and clearer effluent that are desired.

In dual-media filters, it is important that the size of the larger anthracite particles be compatible with the smaller-sized sand particles. When properly chosen, anthracite particles with a 90 percent size will be about three times larger than the sand particles with a 10 percent size. This ratio, called the d_{90}/d_{10} ratio, helps to determine the amount of intermixing at the coal–sand interface. If this ratio is close to two, a sharp interface will be created and excessive head losses and possible mudball formation will result. If this ratio

Table 7-1 Typical filter media design parameters

Media	Media Depth, *in.*	Effective Size, *mm*	Uniformity Coefficient	*L/D* Ratio	Backwash Rate, *gpm/ft²* Winter (0.5°C)	Backwash Rate, *gpm/ft²* Summer (20°C)	Bed Expansion, percent
Sand	24	0.45–0.55	<1.65	1,220	18	25	50
Anthracite/sand	12/16	1.0–1.2/0.45–0.55	<1.4/<1.65	1,120	16.5/18	22.5/25	24/50
GAC/sand	14/16	1.2–1.4/0.40–0.50	<1.4/<1.65	1,180	15/15	21/21	31/49
Deep-bed anthracite	48	0.9–1.1	<1.5	1,220	14	20	22
Deep-bed GAC	48	0.9–1.1	<1.5	1,220	11	15	27

Source: Wolfe and Pizzi 1999.

is close to four, there is too much intermixing at the interface and poor-quality filtrate may result. Consequently, operators should be careful when adding media or "capping off" their filters.

Pretreated water is fed to the filter influent through a pipe or gullet that is designed to cause minimal disturbance. Free-falling or turbulent water should not be allowed to enter the filter because it may disturb the surface of the bed. Baffling may be needed at the influent to reduce any disturbance. Influent piping is designed to provide a velocity of about 2 ft/sec (0.61 m/sec). Effluent and wash-water pipes may be designed for velocities that are two to three times higher than velocities at the influent pipes.

A filter normally includes five valves: influent, effluent, wash-water supply, wash-water drain (sewer), and surface wash or air wash supply. If filter-to-waste capability is added, a sixth valve may be needed. Most valves used on filters are the rubber-seated butterfly type. These valves, which are placed for accessible maintenance, are equipped with position indicators and may be electrically controlled.

Wash-water troughs are used to carry solids away during backwash. By limiting the horizontal distance that floc particles suspended during the wash cycle must travel, properly spaced troughs allow for efficient removal of solids. Troughs must be placed at a level above the filter bed that allows vigorous backwash with minimal loss of media. Engineers usually design the height of a trough to allow for 50 percent bed expansion without loss of media into the troughs. Troughs may be constructed of concrete or fiberglass (Figure 7-6) and should have smooth, level edges to allow for even carryover. Operators should periodically observe the beginning of the wash cycle for uneven wash-water carryover, which indicates a need for maintenance of the troughs.

Underdrains at the bottom of the filter support and retain filter media, collect filtrate, and evenly distribute wash water (Figure 7-7). Unfortunately, underdrains do not always provide a barrier to the finest media in the bed. This small-diameter media can clog the underdrain and lead to uneven filtration or backwashing. These two problems are described here:

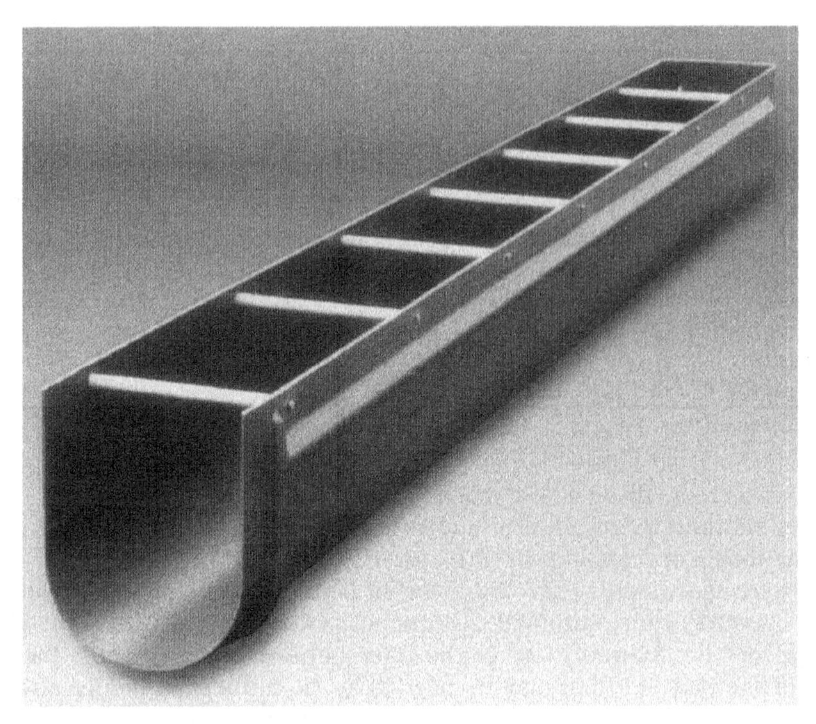

Source: F.B. Leopold, Co. Inc., Zelienople, PA 16063.

Figure 7-6 Fiberglass wash-water trough

- If the filter is operated at design loading rates and an area of the underdrain is clogged, unclogged areas must, by definition, be operating at a higher loading rate. This can cause elevated effluent turbidities. For example, if a 10 ft × 10 ft (3 m × 3 m) bed is designed for a maximum loading of 4 gpm/ft^2 (10 m/hr), it is accepting 400 gpm (1.8 m^3/min) at design. If 15 percent of the underdrain system is clogged, only 85 ft^2 (7.7 m^2) are available. The rest of the system is operating at 4.7 gpm/ft^2 (11.8 m/hr), or 17.5 percent over design capacity.
- That same condition will cause localized "hot spots" at backwash, as the upward flow of wash water must find its way around the clogged areas. As it does, the velocity increases, which can displace and wash away the media. Also, that portion of the bed above the clogged area may not get washed thoroughly.

For these reasons, some filters are equipped with inspection ports at the under-drain level, and operators should periodically take filters off line to perform this important task. Operators should also look for deposits of media underneath the underdrain system, which can indicate a compromised underdrain.

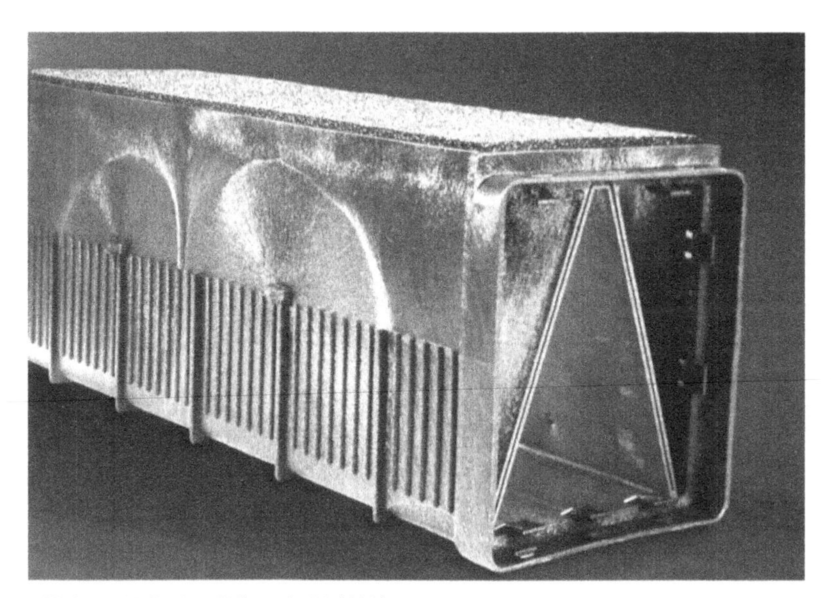

Source: F.B. Leopold, Co. Inc., Zelienople, PA 16063.

Figure 7-7 Underdrain unit with integral media support cap

GRANULAR BED OPERATION AND OPTIMIZATION

The normal cycle for a rapid-sand granular-bed filter begins with placing a clean filter into service and follows the steps of filter ripening, effluent production, and subsequent head loss increases (Figure 7-8).

Eventually, the bed becomes clogged with particulate material, resulting in particle and turbidity breakthrough. At this point, the bed must be cleaned or backwashed. These steps differ in magnitude and frequency from plant to plant, but the process is similar. How these steps are performed can greatly affect finished water quality.

Filter Ripening

When a cleaned filter bed is first put into service, it may not produce effluent of a quality that meets the water treatment plant's goals. Particles in the applied influent water need to attach to the grains of sand or anthracite, otherwise, they will pass into the clearwell with the finished water. Properly treated water (see chapter 5) brings particles that are "sticky," or destabilized, into the filter. These particles attach with a greater efficiency than those that are less sticky. At first, because the bed is so clean, the voids between the grains of sand or anthracite are large and particles have a greater chance of passing through them. As particles are attracted and attached, the bed usually becomes more efficient at attracting more particles: each attached particle becomes an attractive site for more particles to come to rest. The voids get smaller, and it becomes easier for the bed to collect more particles. This process is called *ripening*.

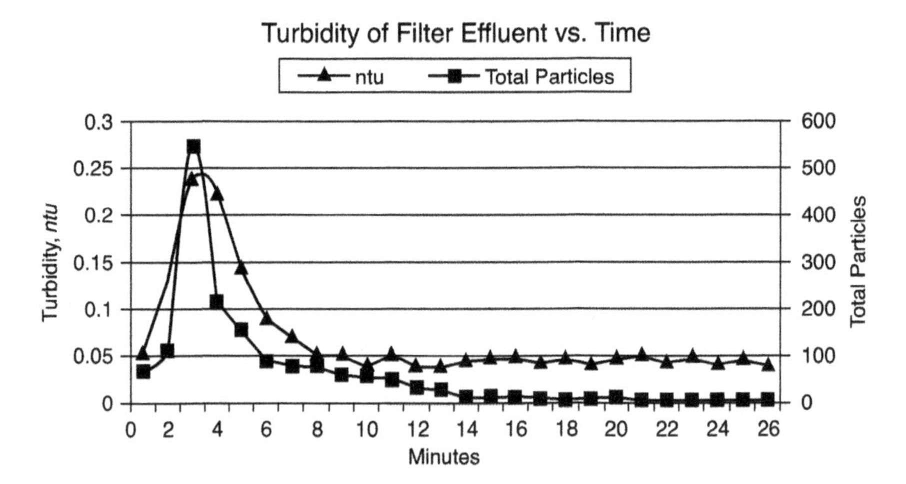

Figure 7-8 Graph of filter effluent turbidity versus time. Note short ripening period for turbidity and particles.

This simplified explanation of the ripening process is helpful for operators because it suggests that some steps are under their control. For example, an operator can see that the better the applied water has been conditioned, the better the effluent will be (usually). Also, it is implied that operating filters at slower rates will improve the attachment process because it will reduce the forces caused by drag through the filter. This is called *fluid drag* and too much of it will shear the attached floc particles. Perhaps the most important lesson is that filters are not just particle-removal devices but are particle-storage devices. Stored particles are in the filter temporarily and must be handled with care.

Head Loss

As the filter ripens and becomes more efficient at producing a clear effluent, head losses increase. Head loss can be defined as a decrease in the available pressure that drives the water through the bed. This loss of pressure is due to frictional pressures brought about by the continual buildup and accumulation (storage) of attached particles. After the filter has been in service for some time, the pressure available from the height of the water in the bed (static head) will equal the frictional pressure and the rate of filtration will be reduced. At this point, breakthrough may occur faster than particles can attach. Beds should be cleaned before this occurs. Operators usually have a maximum head loss at which they will no longer allow the bed to remain in service.

If filters are allowed to operate at head losses that exceed the static head, a vacuum can result. This is called *negative head* and can cause air binding in the filter, i.e., dissolved gases in the water are released, gas bubbles are trapped, and the problem becomes aggravated. Operators of surface water plants often notice air binding in spring when the air temperature is rising but the water is still cold. At these times, operators are often tempted to "bump" the filter to release these trapped gases. Bumping is the act of allowing some wash water to travel through the bed in the hope of reducing head loss and increasing the filter run. This practice should be avoided because it will displace particulate contamination into the clearwell. Air-bound filters should be removed from service and backwashed thoroughly, regardless of the length of the run. A better practice is to allow the water level on top of the bed to remain at a height that will minimize head loss increases and therefore minimize air releases.

Rate-of-Flow Control

The ability to control the rate of flow through filters is important. Plants with multiple filters may not distribute the total plant flow in a reasonably equal manner due to varying head losses among the filters. Also, without flow control, filters may accept sudden changes (increases) in flow, which can cause poor filtrate quality. In general, flow is controlled in one of two ways: mechanically or nonmechanically (where hydraulics dictates the flow). Rate-of-flow control may be better understood by classifying the filters as either "equal-rate" or "declining-rate" filters. The concept of rate control is complex and will not be discussed in great detail. What follows is a general overview of the subject.

Equal-rate flow control is accomplished by variable-level influent flow splitting, proportional-level influent flow splitting, or proportional-level equal rate control. None of these conditions can be classified as "constant-rate filtration" because they can only occur when the total plant flow rate and the number of filters in service are held constant. Removing a filter from service usually places a burden on other filters, and so constant rate over time is rarely seen.

Variable- and proportional-level flow splitting are attained by placing weirs or orifices on each filter inlet above the filter's maximum water level. The filter effluent discharges to the clearwell at a level above the surface of the filter medium. As solids accumulate in the bed, the water level rises in the filter box to provide the head that is needed to drive the flow through the filter. The water level in each box is different and depends on the amount of clogging. When the level reaches a predetermined height, the filter must be backwashed. No instrumentation is used to control flow rate or head loss measurement. Operators can view relative head loss by looking at the level in each box.

Influent flow-splitting weirs are also used for proportional-level operation, which is accomplished mechanically. Each filter requires a level transmitter and controller, with a modulating valve on the effluent. This valve does not control flow—it controls the level on top of the filter. As the water level on the filter changes, the valve attempts to compensate. As head losses accumulate, the valve cannot continue to operate and so it shuts down, indicating a need for washing.

The most common proportional-level system is the mechanical-control type. This system splits the flow equally among all filters in operation by keeping the influent channel level relatively constant. Effluent valves are modulated to control flow rate, and filters are washed at a predetermined head loss.

Variable- and proportional-level systems are also found in declining-rate filters. Because the influent enters each filter at a level below the normal water level in each filter and discharges to the clearwell at a level above the filter medium, the filters operate with the same available head at any instant. The cleanest filters will therefore operate at the highest flows, and the dirty filters will operate at lower flow rates. All filter rates decline in a step-wise fashion after each backwash. When a clean filter is placed into service, it assumes the highest flow. All other filters in service step down from that point.

Granular Bed Maintenance

Granular filtration beds tend to be maintenance-intensive. It is common for operators to ignore these important maintenance tasks, which can lead to filter problems that jeopardize filtrate quality. As these problems become more severe, they also become more expensive and difficult to remedy. Most maintenance tasks can be accomplished with the simple tools found in most water treatment plants (Figure 7-9).

Bed Depth

Because filter backwashing and other operating practices can result in media loss in the filter, the total media depth should be measured periodically to determine if the existing media still meet original specifications. It is not unusual to lose some media each year, especially if the bed is of dual-media construction comprised of sand and anthracite. The backwash velocities required to clean a sand bed may wash away some of the anthracite, which has a density less than that of silica sand. Excessive media loss can, however, indicate poor operational technique or filter bottom problems and will eventually result in filter effluent degradation.

How deep is deep? In part, a filter's ability to trap floc particles in suspension at normal operating rates is a function of bed depth and media size. The minimum standard for proper filter construction is that the ratio of the bed depth, in millimetres, divided by the effective size, in millimetres, is equal to or greater than 1,000, or

$$l/d_e \geq 1{,}000$$

where l = depth of the filter bed, in millimetres, and d_e = effective size of the media.

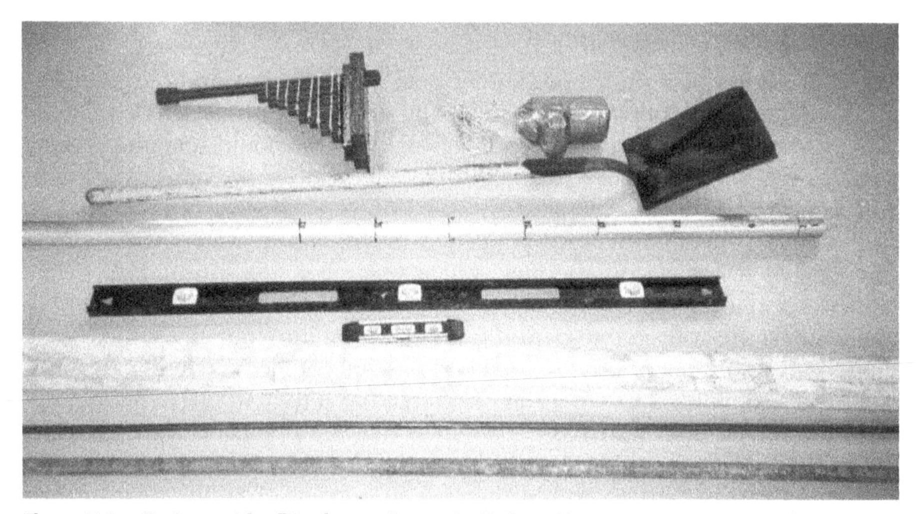

Figure 7-9 Tools used for filter inspection typically found in any water treatment plant

In practice, it is common for plant operators to maintain a bed that is 15 percent in excess of the minimum standard. For example, a 6-ft-deep monomedia bed comprised of 1-mm anthracite coal would be more than sufficient because 6 ft is 72 in., and 72 in. × 25.4 mm/in. = about 1,830 mm. Using the formula, 1,830 mm/1 mm = 1,830—a ratio that meets the guidelines for sufficiency.

Beds other than monomedia or dual-media, such as tri-media or coarse monomedia beds, may require a higher ratio.

How does an operator measure bed depth? Bed depth is a measure of the amount of filter media in the cell and does not include the support gravel or underdrains. After checking the original specifications, measure the bed depth by poking a ⅜-in. steel rod into the media until the gravel or support media is reached. Experienced operators develop a "feel" for the difference in resistance that the rod meets at the interface of the media and the gravel. Others can hear the distinctive "crunch" as the rod hits its mark. When the depth has been found, the rod should be pinched at the surface and carefully pulled out. The rod should be taped at that mark and measured in inches. At this point, the rod should be poked into other areas of the filter bed to determine if the filter media is level or the underdrain is disturbed. Within reason, the level should always come close to the tape mark wherever the rod is poked in.

This measurement is very important when discussing bed expansion, and it will quickly indicate when additional filter media need to be installed. Take care when replacing filter media—adding media of a different specification to an existing bed may cause filtering problems over time. Also, be sure to measure the depth of each type of media in dual- and mixed-media beds. It is not uncommon to lose sand in a dual-media bed. Adding anthracite to a bed that has lost sand may worsen the problem.

Bed Expansion Measurement

The cleanliness of a filter bed depends on an operator's ability to achieve and sustain a proper bed expansion during the backwash cycle. This expansion, commonly recorded as percent of bed depth, should remain constant year round. Water supplies that are subject to wide temperature variations will need to use a different wash-water flow rate to keep this constancy. In general, a higher flow rate is needed during warm water periods than during cold water periods because warm water has less lifting power than cold water due to the difference in viscosity. Therefore, operators need to make adjustments from one period to the next. Failure to make these adjustments can result in a poorly cleaned bed or a loss of media due to carryover.

As an example, consider a filter bed that has a total depth of 30 in. Suppose the operational staff has determined that a 30 percent bed expansion keeps the bed clean and able to produce the desired effluent turbidity. The calculated bed expansion would then be 30 percent of 30 in., or 9 in. Any wash-water rate of flow that achieves a 9-in. expansion would be acceptable to the plant staff. The operator simply needs to periodically measure this expansion and adjust the flow rate to maintain the 9-in. expansion.

Operators have devised simple yet effective tools for this purpose, chief of which are the "pan pipe" expansion tool and the circular disc tool (see Figure 7-10). The expansion tool is a series of plastic pipes mounted to a stand at 1-in. intervals. When secured to the top of the bed with cable ties, backwash expansion can be measured by observing the anthracite that has been displaced into the pipes. The highest pipe that has media in it will give an indication of the height of the expansion.

The circular disc tool is simply a white plastic disc fastened to a rod that is marked in inches. The tool is placed so that the disc sits at the top of the bed. A measurement is then taken and recorded. The tool is removed, and the filter is backwashed. At the height of the backwash, the rod is lowered into the washing filter until the disc "disappears" just at the top of the expanded anthracite. A reading in inches is taken at that point. Subtracting the smaller reading from the larger reading gives inches of expansion.

Both methods are usually more reliable than the older method of measuring "rate of rise" by watching the backwash water level travel from one point to another. That method requires the operator to drop the bed level to a predetermined mark and then record the time it takes to reach the next mark. This reading, which is inches of rise, can be converted to gallons per minute per square foot by dividing by 1.6. It does not reveal the extent of bed displacement but rather reveals the backwash water rate.

Filter Coring and Solids Retention

The effectiveness of the backwash scheme and the efficiency of the filtering process can be determined by taking samples, or cores, of the bed at successive depths and measuring solids retention. Filter coring is an inspection technique that examines the amount of floc particles that are attached to

Figure 7-10 Filter bed expansion tool. Pipes are arranged in 1-in. increments.

the filter media at successive levels for the entire depth of the bed. Solids retention data are expressed as nephelometric turbidity units per 100 g of media. If a solids retention profile is constructed for each filter at a time when the filter is consistently producing acceptable finished water under normal operating conditions, that profile can act as a "baseline," or description of the filter against which other profiles may be compared (see Table 7-2). The table suggests the ranges of floc retention amounts and corresponding actions that might be followed.

Note that the solids retention guidelines are a reference for plants that have not performed this analysis previously. These guidelines may not be appropriate for each filter; therefore, the operator is advised to use these guidelines only as intended. Examples are given in the following paragraphs.

How to perform a solids retention analysis. The tools for extracting a core sample for solids analysis are easily made at most water treatment plants. A simple tool (Figure 7-11) made of a 5-ft length of ½-in. electrical conduit is used as the corer. The tool should be marked with an indelible marker at the 2-in., 6-in., 12-in., 18-in., 24-in., and 30-in. marks, etc. (i.e., for the entire length necessary to reach the bottom of the bed). This configuration of tool material and diameter seems to work best. Other materials (plastic pipe) and diameters do not seem to provide sufficient adhesion for both sand and anthracite, nor do they provide volume sufficient for sample analysis. In addition to the coring tool, sealable plastic bags are needed to hold

Table 7-2 Solids retention guidelines for core samples taken after washing

Solids Retained ntu/100 g media	Condition of Bed	Action Needed
<30	Too clean	Examine wash rate and length
30–60	Well cleaned and ripened	No action needed
60–120	Slightly dirty	Reschedule retention analysis soon
>120	Dirty	Reevaluate filter wash system and procedures
>300	Mudball problem	Rehabilitate bed

Source: Wolfe and Pizzi 1999.

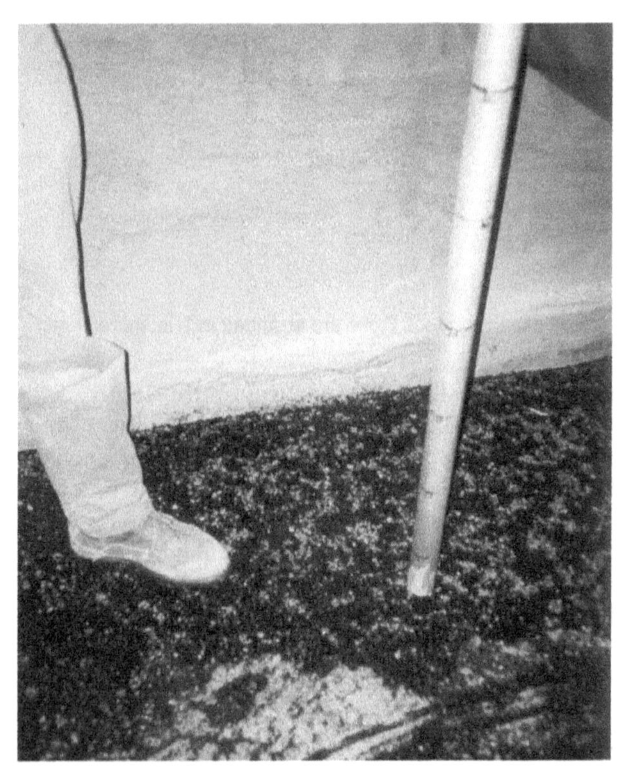

Source: Cleveland, Ohio, Division of Water.

Figure 7-11 Operator using coring tool. Note markings on pipe, which correspond to layers in the bed; 1½-in. electrical conduit is used.

the media samples. Clear plastic bags allow for visual inspection of the media, making it possible to observe mudballs and other debris that may be present. A filter core set should be collected from a filter bed upon completion of an entire filter run. This set of samples is the "before" set because it is taken before the filter is backwashed. After the filter is backwashed, another set of filter core samples is taken. This is the "after" set because it was taken after completion of a backwash. A before-and-after set of bags should be provided, two each for each stratum (i.e., 0–2 in., 2–6 in., 6–12 in., etc.) that will be sampled. The bags should be coded with the marker to correspond to the levels on the coring tool (Figure 7-12).

Coring is performed after the bed has been drained. For best results, begin the procedure while the bed is still damp—also this will provide for optimum media adhesiveness. Place two small (2 ft × 2 ft) pieces of plywood on the bed so that two operators can stand comfortably on the filter media with minimal bed disruption. One operator inserts the coring tool into the bed at right angles until the 2-in. mark is reached (i.e., to obtain the media sample for the 0–2 in. strata). Carefully extracting the tool so as not to lose the sample, the operator empties the sample from the coring tool (by blowing into the other end of the conduit) into the appropriately marked bag, which is held open by the second operator. This procedure is repeated two more times at two nearby, but separate, locations so that reasonably representative samples of the upper 2 in. of media are obtained.

The next extraction is obtained from the 2-in. to 6-in. strata, and the operator must carefully insert the coring tool into the same three holes to get these samples. Care must be taken so as not to contaminate the coring tool with material that may have fallen into each sample hole. The tool exterior should be wiped clean between each sampling as an added precaution. This procedure is repeated until samples from all representative strata of the bed are obtained. As each appropriately marked bag is filled, it is sealed and stored for future analysis. For dual- and mixed-media configurations, at least one bag will yield a mixed layer (e.g., sand and anthracite), which should be closely examined for any interesting characteristics, including mudballs, particle size distribution, and, of course, the depth at which it was obtained. The d_{90}/d_{10} ratio of the two media has a direct impact on the length of this stratum.

Figure 7-12 Lab setup for solids retention analysis. Note plastic bags with core samples.

After all "before" core extractions have been obtained, the bed should be backwashed in the usual fashion. This is an excellent opportunity to observe the backwash process for any anomalies. Additionally, several samples of the backwash water from the point at which the water exits the trough can be obtained for turbidity analysis. A plot of backwash water turbidity versus time can offer additional insights (i.e., at which point during the backwash does the turbidity of the backwash water reach its minimum level?). After backwashing, "after-core" samples of the bed should be obtained by repeating the entire filter-coring procedure.

The bags containing the before- and after-core samples are brought to the lab for turbidity, or floc retention, analysis. (This is also a good time to visually check each bag for mudballs.) The goal is to measure floc retention in terms of turbidity, as nephelometric turbidity units per 100 g of media. In the lab, 50-g portions from each bag are weighed and placed into separate 500-mL beakers. Five successive 100-mL portions of turbidity-free water are used to wash down and agitate each media sample for 30 seconds. After each sample washing, the 100-mL portions are carefully poured into a 1,000-mL beaker and turbidity of the resultant suspension is measured. When all measurements are recorded, the results are multiplied by 2 (i.e., to obtain turbidity in terms of nephelometric turbidity units per 100 g of media) and graphed. See Figure 7-13 for an example of the plot.

As previously mentioned, suggested guidelines for sludge retention profiles for filters are given in Table 7-2. Operators should remember that conditions at their utility (i.e., process type, operating habits, etc.) may produce significant deviation from these published guidelines. Consequently, these guidelines may be inappropriate or misleading at an individual utility and are provided to encourage vigorous examination of local filter operating and backwashing conditions. In general, if filter effluent quality is good, the guidelines in Table 7-2 may be considered irrelevant with respect to local conditions. The operator can simply analyze for solids retention and develop his or her own guidelines as a baseline. If filter effluent quality is poor, these guidelines offer a good starting point and a realistic goal for comparison. Each operator should, in time, be able to develop a "normal" set of guidelines for use based on repeated and regular analysis of the utility's filter conditions.

Filter Backwashing

The most common problems associated with properly designed filters are those induced by the operator. The backwash procedures that operators do or do not follow will determine whether the filter produces quality effluent and can lead to early filter degradation. For this reason, it is important that each plant establish clearly written backwash procedures and that all operators follow them as written. It is not uncommon to see water treatment plant operators in the same plant using different backwash procedures, with all operators convinced that their method is the proper one.

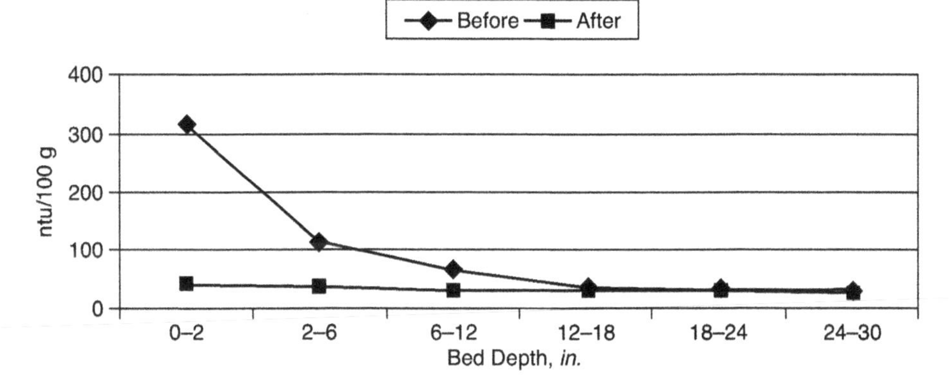

Figure 7-13 Graph of solids retention. Graph shows that the filter bed was not overly cleaned. This bed should ripen quickly when put back into service.

Because there are many design differences among filters and because each water treatment plant is unique, a single backwashing procedure cannot be prescribed. Each plant must develop its own procedures based on design recommendations and experience. The following are some general filter backwash recommendations to be considered at each plant:

- If surface washers are available, use them before the start of the backwash and continue using them as the backwash rate is ramped up to a maximum (Figure 7-14). Be careful not to allow the surface washers to contribute to media loss.
- Surface washing can be a great help in preventing the formation of mudballs. Once mudballs form and continue to grow in size, they can ruin the ability of a filter to produce low-turbidity water. Mudballs that grow in size also grow in weight and will sink into the filter and go undetected.
- It may be advisable to ramp up the wash-water rate in stages, especially if there is more than one type of media in the bed (dual and mixed beds), a ramped wash-water rate that expands each bed properly is advisable. For example, a typical dual-media bed of anthracite and sand requires two rates. The initial rate expands and cleans the anthracite but will not sufficiently clean the sand. However, if the operator ramps quickly to the high rate to clean the sand, the anthracite will be overly expanded and not cleaned properly. Each type of media has a different specific gravity and therefore will require different wash-water rates for expansion. Refer to Table 7-1 for guidelines.
- Do not wash the filter too much. Most operators backwash until the turbidity of the wash water coming through the bed is less than 10 ntu or until they can see the surface washers or other piping that is above the bed. Washing the filter too long will remove the necessary bed ripening; the filter will then have to be reripened for a longer period

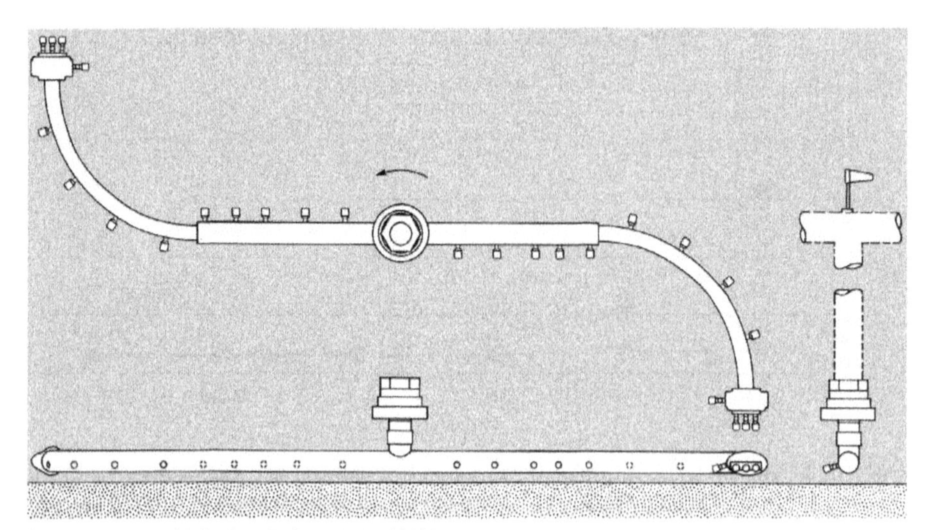

Source: F.B. Leopold, Co. Inc., Zelienople, PA 16063.

Figure 7-14 Surface agitator. Operators should check nozzles periodically for plugging.

of time before it is useful. This also wastes wash-water pumping. It is normal for operators to want to do a thorough job when they clean the filter, but too much of a good thing can do harm. Operators should periodically collect samples of the backwash water at 1-min intervals for the length of the wash and analyze them for turbidity (Figure 7-15). Figure 7-16 displays filter wash turbidity over time. Note that after 6 min, the backwash no longer produced appreciable solids, yet the operator needlessly continued the wash.

- Wash at a rate that will expand the bed properly under all wash-water temperature conditions. Because colder water has more lifting power than warmer water, it is usually necessary to adjust the wash-water rate from summer to winter for most surface water plants in the northern regions. As previously mentioned, the percent bed expansion should not change from season to season, but the wash-water rate needed to achieve the expansion may change.

An insufficient wash-water rate will not properly clean a filter no matter how long it is washed. The turbidity of the wash water coming through the filter may lessen, leading the operator to believe that the bed has been cleaned, but this may not be so. This practice usually leads to mudball formation. Washing a filter at 15 gpm/ft^2 for 10 min requires the same amount of water as washing the bed at 7.5 gpm/ft^2 for 20 min, but the results will be different. Filters get cleaned by using the correct amount of wash water at the correct wash-water rate for the correct amount of time. That correctness is best determined at each individual plant using procedures outlined earlier. Table 7-3 shows temperature correction factors for backwash rate values if the rate at 25°C is known.

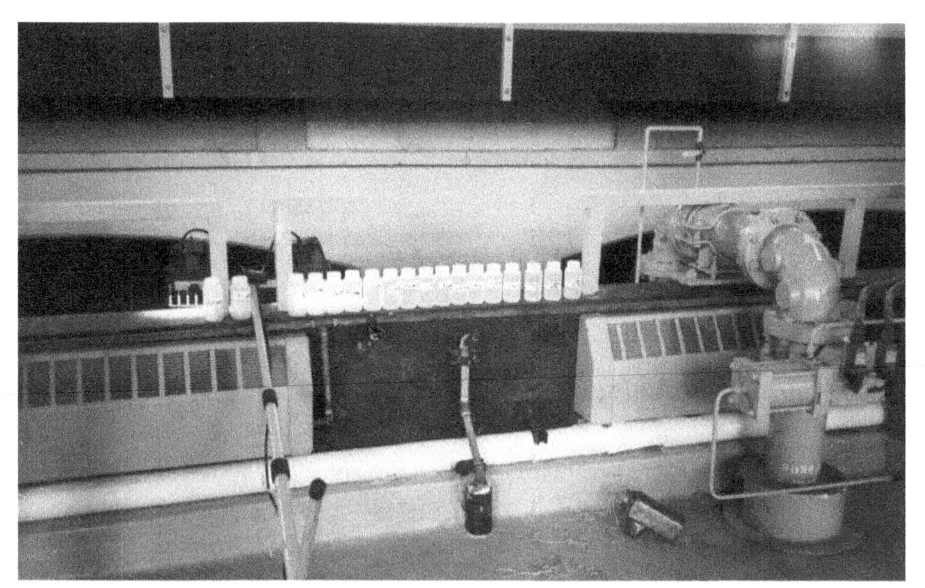

Source: Cleveland, Ohio, Division of Water.

Figure 7-15 Bottles prepped for wash-water sampling. Each bottle is labeled in 1-min increments. Use plastic bottles to avoid breakage.

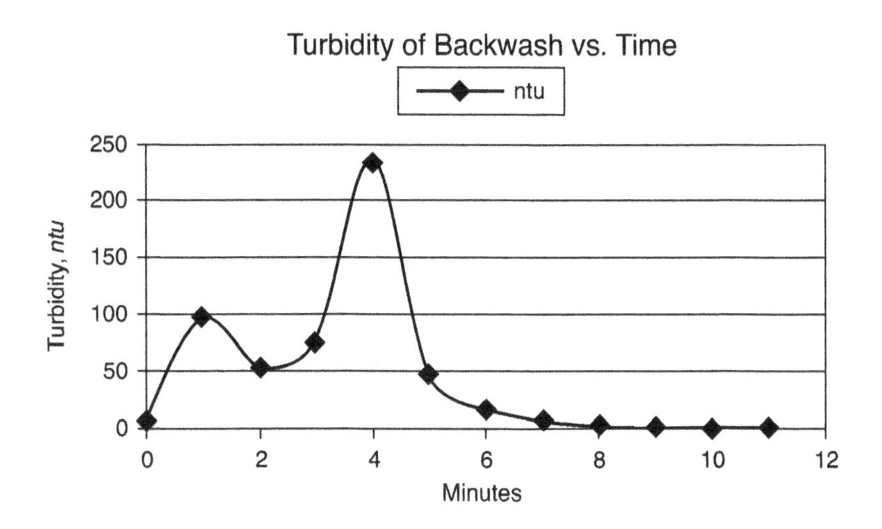

Figure 7-16 Graph of backwash turbidity versus time. This bed appeared to be clean in 6 min, but was washed longer than necessary. This practice may waste wash water.

Table 7-3 Backwash rate temperature correction factors

Temperature, °C	Multiply 25°C Value by
30	1.09
25	1.00
20	0.91
15	0.83
10	0.75
5	0.68

Source: MacPhee and Becker 1999.

Example: If 5,000 gpm provides a suitable backwash at 25°C, an operator could expect to achieve the same bed expansion at 10°C by using 3,750 gpm (0.75 × 5,000).

- If air scour is used prior to (not concurrent with) water backwash, do not assume that less wash-water rate can be used. Air scour will do a good job of knocking the adhered floc particles loose from the filter media, but they still have to be swept up from the bed and into the troughs for removal. This probably takes as much wash-water rate (gallons per minute per square foot) to accomplish as those filters without air scour; however, it may not take as much wash-water volume.
- A review of operational procedures for backwash at most water treatment plants that use conventional granular beds reveals that operators usually use 100–150 gal of wash water per square foot of surface media and, on average, use no more than four percent of the plant production to backwash filters.
- After backwash, leave the filter off-line or "rest" it if possible. Many operators report that a rested filter ripens more quickly when put online than will a filter that is washed and put immediately into service. This practice of resting filters also allows the operational staff to have a fresh filter in reserve that can be put online when another filter is removed for washing. This minimizes hydraulic shock in rate-controlled filtration schemes.
- An operator cannot perform a proper backwash without visually inspecting the process as it occurs. With newer, sophisticated, and automated backwash systems, it is tempting for the operator to control the sequence from a remote location (Figure 7-17). However, the operator cannot see and tend to problems that may be occurring at the backwash site. Operators should observe each wash event for uneven surface agitation and expansion, for "hot spots," and for changes in backwash water turbidity. During the wash, the operator should hose down the side walls of the filter box and the associated piping to remove any accumulated scums, which can harbor pathogens.
- Think about how the filter wash event will affect all other plant operations such as pumping rates, wash-water return storage availability, maintenance, and any other planned activities. Operators should know

Source: Hach Company.

Figure 7-17 Operator preparing for filter backwash

in advance when they will be washing a filter and what may or may not happen when they do.

Operational Considerations

Through experience gained over time, operators learn techniques that are helpful for operating filters in their plant. What follows are general operational techniques used by most operators.

When filter effluent turbidities begin to increase above what is normal, the problem can usually be traced to one of two areas: the filter(s) or pretreatment (Figure 7-18). This may seem simple but it is often misdiagnosed, and operator actions sometimes worsen the problem. The best designed and operated filters usually will not produce good-quality effluent if pretreatment is poor. It is instinctive for most operators to bring extra filters online when the pretreated water has deteriorated, but this may make things worse. Unripened filters, in particular, will not handle poorly treated influent and may actually pass higher turbidities when brought online under these conditions. Of course, if the elevated turbidities are due to compromised filters or if the filtration rate is too excessive, it may be wise to use other filters if available.

Source: Hach Company.

Figure 7-18 Online turbidimeter

Operators should not be fooled into thinking that longer filter run lengths are more efficient. If filter washes are accomplished with 100 to 200 gal/ ft^2/wash (4 to 8 m^3/m^2/wash) and backwash downtime is about 30 min, no more efficiency will be gained after a unit filter run volume of 5,000 gal/ ft^2/run (200 m^3/m^2/run). This is true over a wide range of filtration rates. Longer filter runs will increase the risk that larger amounts of particles will empty into the clearwell if the filter is upset. All other considerations being equal (water quality, filter rate, etc.), the longer the filter run, the larger the amount of particulates is being stored in the filter.

Filter Run Profile

Operators should periodically examine the effluent turbidity or particle count data of each filter and create a graphic profile of this data. An example of a typical filter profile is shown in Figure 7-19. The profile should show the output of a normal filter run from startup through the end of the cycle. The data will show the ripening characteristics of the filter bed and the effect of hydraulic and treatment changes that occur during the period. Operators should be able to explain any spikes found in the data. These types of profiles are required in the filter exceptions reporting provisions of the Interim Enhanced Surface Water Treatment Rule (IESWTR).

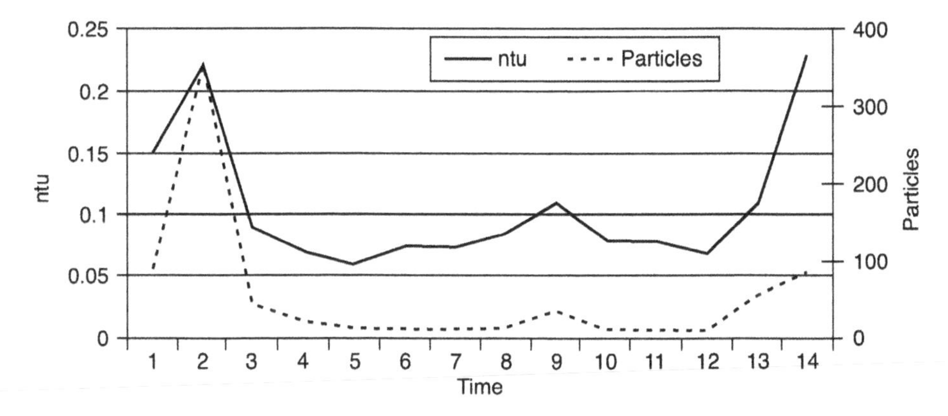

Figure 7-19 Filter run profile. Note ripening period and subsequent breakthrough at end of filter run. Also note that spiking occurred, possibly due to hydraulic change at time "9."

Importance of Filtration

With promulgation of the IESWTR, the water quality at a surface water filtration plant is judged by the output of the worst filter online. It is no longer determined simply by the quality of the combined filter effluent, which is an average of all filters. This regulatory philosophy returns water plant operations to a time when filtration was the most important unit process available in the quest for public health. Before the advent of chlorination, operators depended on filtration to remove harmful microorganisms from the water supply. Public records of many cities reveal that when filtration was installed at water plants, a decrease in community disease was noticed. This direct cause-and-effect relationship between unit process and public health was the single most important discovery of the day—a shift in thought that brought science into the craft of water supply and treatment.

Soon after, chlorination was discovered and brought into vogue as a first line of defense against disease-causing organisms. Generations of plant operators lost sight of the importance of filtration and began to think of it as a simple polishing process, one that improved the "clarity" of the water. Improving the clarity, it was said, would discourage consumers from seeking alternative and perhaps less safe water supplies.

Improved detection techniques have shown that protozoa in water supplies can sometimes be quite numerous, both in nature and in the plant process—and in water that is crystal clear. It is known that chlorination cannot inactivate many of these organisms. Other unit processes must be optimized to accomplish that. Optimized filtration, in concert with chlorination, serves as a synergistic and scientific approach to that end and contributes to the overall improvement of community health.

BIBLIOGRAPHY

AWWA. 2009. M30—*Precoat Filtration*. Denver, Colo.: American Water Works Association.

Bellamy, W.D., G.P. Silverman, D.W. Hendricks, and G.S. Logsdon. 1985. Removing *Giardia* Cysts With Slow Sand Filtration. *Jour. AWWA*, 77(2):52–60.

Beverly, R.P. 2005. *Filter Troubleshooting and Design*. Denver, Colo.: American Water Works Association.

Cleasby, J.L., D.J. Hilmoe, C.J. Dimitracopoulos, and L.M. Diaz-Bossio. 1984. Effective Filtration Methods for Small Water Supplies. Final Report. USEPA Cooperative Agreement CR808837-01-0. NTIS No. PB84-187-905.

Edzwald, J.K., ed. 2011. *Water Quality and Treatment*, 6th ed. New York: American Water Works Association and McGraw-Hill.

Kawamura, S. 2000. *Integrated Design of Water Treatment Facilities*, New York: John Wiley & Sons Inc.

Kirner, J.C., J.D. Littler, and L.A. Angelo. 1978. A Waterborne Disease Outbreak of Giardiasis in Camus, Wash. *Jour. AWWA*, 70(1):41.

MacPhee, M.J., and W. Becker. 1999. Filter Backwashing: When and How. *Opflow*, 25(3):10–12.

Pizzi, N.G. 1996. Optimizing Your Plant's Filter Performance. *Opflow*, May, p. 1.

Pizzi, N.G. 2000. Water Treated Right at Oak Creek Plant. *Opflow*, September, p. 6.

Pizzi, N.G., R.O. Schwarzwalder, and J.J. Sadzewicz. 1997. Using Negotiated Guidelines in a Facility Upgrade. *Jour. AWWA*, 89(3):89–95.

Wolfe, T.A., and N.G. Pizzi. 1999. Optimizing Filter Performance. *Jour. NEWWA*, 113:6–21.

Chapter 8

Disinfection

The primary goal of disinfection is to reduce the numbers of pathogenic microorganisms in water. Other treatment processes may contribute to pathogen reduction; however, that is not their primary purpose. There are numerous types of disinfection, all of which employ chemical or physical agents.

Chemical disinfection through the use of chlorine has existed for more than a century. In early times, chlorine disinfection helped eradicate, or at least greatly reduce, the incidence of classic waterborne diseases such as typhoid and cholera. Disinfection with chlorine is probably one of the most important engineering accomplishments with respect to public health. In recent years, hardier organisms have emerged (e.g., *Cryptosporidium*) that resist traditional disinfection methods, and other technologies to augment or replace chlorine disinfection are being studied. The traditional and the more modern disinfection practices are discussed in this chapter.

BASICS OF CHEMICAL DISINFECTION

There are some fundamental concepts that operators need to know about chemical disinfection. When an oxidant such as chlorine is applied to water, it reacts with many of the impurities in the water to form compounds, some of which do not have the power to disinfect. Because the chlorine is used up in these reactions, it is said that the water has a "demand" for chlorine. Of course, some of the chlorine is used up destroying the pathogenic organisms, which also contributes to the demand for chlorine. Other impurities that use up chlorine include ammonia, iron and manganese, hydrogen sulfide, and organic material, including organic nitrogen. As long as these impurities are found in the source water, some of the disinfectant will be used up in the treatment process. Some reactions, like that of chlorine and ammonia, take place quickly. Others, like the reactions between chlorine and organic material, may take longer.

The oxidation of soluble inorganic contaminants must take place before filtration because the oxidized form of some constituents will stain fixtures if allowed to enter the distribution system. Additionally, all inorganics will create an oxidant demand. Table 8-1 shows the amount of chlorine needed to oxidize some of these inorganic compounds. It should be noted that the process also consumes alkalinity.

Most disinfection strategies require that an excess of disinfectant remain in the water as it travels through the distribution system. This excess, or residual, is desirable because it helps to continue the disinfection process, as well as provides added protection from unforeseen contamination or slime growth in the distribution system. When all of the demand for disinfectant is met, any added amount of chemical will produce this residual.

Effective disinfection takes place when the needed amount of disinfectant comes in contact with the microorganisms in the water and has sufficient time to disrupt the normal life processes of the organism. Pathogens such as viruses and *Giardia lamblia* have minimum requirements for disinfection concentration and contact time, and these requirements are temperature and pH dependent. The concept of disinfection sufficiency is represented by the equation $C \times T$, where C represents a known concentration of disinfectant, in milligrams per liter, that is multiplied by T, the time in minutes. Achieving the appropriate $C \times T$ value, in mg/L-min, is a requirement of the Surface Water Treatment Rule (SWTR) and is explained later in this chapter and also in chapter 1.

Practical Aspects of Chlorination

Disinfection using chlorine, or chlorination, is usually accomplished by adding gaseous chlorine, liquid sodium hypochlorite, or solid calcium hypochlorite (Figure 8-1). The application of any of these chemicals results in "available chlorine," or hypochlorite, and is represented by the formula OCl^{-1}. Gaseous chlorine is considered pure, and so 100 percent of it goes into producing hypochlorite. Liquid sodium hypochlorite is purchased or generated at the site of application. It has a working percentage of 1 to 16 percent available hypochlorite, which must be factored by using mathematical formulas such as those found in appendix A. Granular calcium hypochlorite

Table 8-1 Chlorine requirements and alkalinity consumption with inorganics

Inorganic Reactant	Part of Chlorine Required per Part of Inorganic	Alkalinity Consumed per Part of Chlorine Added
Iron	0.6	0.9
Manganese	1.3	1.5
Nitrite	1.5	1.8
Sulfide to sulfur	2.1	2.6
Sulfur to sulfate	6.2	7.4

Source: Connell 1996.

Source: Siemens Industry Inc.

Figure 8-1 Gas chlorinator

also is not pure, available chlorine; generally, it is found to be 65 to 70 percent calcium hypochlorite. Operators should note the effect that chlorination has on pH. Gaseous chlorine tends to lower pH, while the hypochlorite compounds tend to raise it. The degree to which the pH is affected is dependent on the water's buffering capacity. Low-alkalinity waters can exhibit very noticeable swings in pH when chlorinated.

Properties of Chlorine

Chlorine is a greenish yellow gas with a penetrating and distinctive odor, and it is very harmful to humans, even at low concentrations in air. The gas is 2.5 times heavier than air. For this reason, it is necessary to move to high ground when a chlorine leak occurs (Figure 8-2). Chlorine gas has a high coefficient of expansion, meaning that it expands easily and with great volume when heated. This latter property makes it dangerous because it could easily burst its container if heat is applied. Chlorine will not burn but it will support combustion. Also, water makes chlorine gas corrosive to steel. For these reasons, operators should follow these commonsense rules when using chlorine:

- Always use a self-contained, positive-pressure breathing apparatus when working on chlorine cylinders.
- Never work on cylinders alone.
- Never apply heat to a chlorine cylinder.
- Never apply water to a leaking or burning cylinder, because the leak will grow worse.

Source: Conneaut, Ohio, Water Department.

Figure 8-2 Installation of 150-lb chlorine cylinders. Tandem unit allows for dual withdrawal.

Chlorine Reactions

When chlorine is added to water, several chemical reactions take place. Some involve the water molecules and some involve organic and inorganic substances in the water. To study these reactions, it is necessary to become familiar with some terms that are commonly used in the practice of chlorination. Small amounts of chlorine added to water will combine with organic and inorganic materials to form chlorine compounds. As more chlorine is added, a point is reached where the reaction with these organic and inorganic materials stops. At this point, the *chlorine demand* has been satisfied.

The chemical reactions that took place in satisfying the chlorine demand created compounds, some of which have disinfecting properties and some of which do not. In the same fashion, chlorine will react with water molecules to produce compounds that have disinfecting properties. The total of all the compounds with disinfecting properties, plus any remaining uncombined chlorine, is known as *chlorine residual*. The presence of chlorine residual is measurable and is an indication that chemical reactions have taken place and that there is still some *free* or *combined available residual* to kill the microorganisms that may be introduced to the water.

The amount of chlorine needed should be added together to satisfy the chlorine demand and the amount of chlorine residual needed for disinfection to arrive at the *chlorine dosage*. This is the amount of chlorine that must be added to the water to disinfect it.

Chlorine Dosage Equation

Chlorine Dosage, mg/L = Chlorine Demand, mg/L + Total Chlorine Residual, mg/L

Chlorine Residual Equation

Total Chlorine Residual, mg/L = Combined Chlorine, mg/L + Free Chlorine, mg/L

Reaction With Water

Free chlorine combines with water to form hypochlorous acid (HOCl) and hydrochloric acid (HCl):

$$Cl_2 + H_2O \leftrightarrow HOCl + HCl$$

Depending on the pH of the water, hypochlorous acid may be present in the water as the hydrogen ion and hypochlorite ion dissociate.

Hypochlorous acid dissociation is represented by the following equation:

$$HOCl \leftrightarrow H^+ + OCl^-$$

In dilute solutions of chlorine in water (such as those found in a water treatment plant), the formation of HOCl is complete and leaves little free chlorine (Cl_2). Hypochlorous acid is weak and poorly dissociates at pH levels below 6. Thus, any free chlorine or hypochlorite (OCl^-) added to the water will immediately form either HOCl or OCl^-; the species formed is thereby controlled by the pH of the water. This is extremely important because HOCl and OCl^- differ in disinfection strength. HOCl has a much greater disinfection potential than OCl^-. Normally, 50 percent of the chlorine present in water with a pH of 7.3 will be in the form of HOCl and 50 percent will be in the form of OCl^-. The higher the pH level, the higher the percent of OCl^- and therefore the less disinfecting ability.

Chlorination With Hypochlorite

The use of hypochlorite compounds to treat water achieves the same results as chlorine gas in that they each produce HOCl in water as a disinfectant. Hypochlorite may be applied in the form of calcium hypochlorite, $Ca(OCl)_2$ (commonly known as HTH), or sodium hypochlorite, NaOCl (referred to as bleach in liquid form). The chemical reactions of hypochlorite in water are similar to those of chlorine gas.

Calcium Hypochlorite

Calcium Hypochlorite		Water		Hypochlorous Acid		Calcium Hydroxide
$Ca(OCl)_2$	+	$2H_2O$	$\rightarrow$	$2HOCl$	+	$Ca(OH)_2$

Sodium Hypochlorite

Sodium Hypochlorite		Water		Hypochlorous Acid		Sodium Hydroxide
$NaOCl$	+	H_2O	$\rightarrow$	$HOCl$	+	$NaOH$

When chlorine gas is used for disinfection, 100 percent of the gas is available as chlorine. Therefore, if the chlorine demand and the desired residual require 50 lb of chlorine to be fed, the chlorinator would be set at exactly 50 lb/day. However, when hypochlorites are used, the compounds contain only a percentage of the available chlorine that the gas does. HTH is about 65 percent available chlorine, and sodium hypochlorite is usually found to be 5 to 15 percent available chlorine. Because of this, more pounds per day must be fed to obtain the same results that would be expected from chlorine gas. To calculate the pounds-per-day requirement of hypochlorite, first determine the pounds per day of chlorine gas needed, then divide that number by the percentage of available chlorine in the hypochlorite. The calculation is

lb/day hypochlorite = (million gallons per day)(8.34)(mg/L gas)/
percent available chlorine in hypochlorite

Example 1: A total chlorine dose of 12 mg/L is needed to treat a source water. If the flow is 1.2 mgd and the hypochlorite has 65 percent available chlorine, how many pounds per day should be fed?

Answer: 12 mg/L × 1.2 mgd × 8.34/0.65 = 185 lb/d HTH.

Example: 2: A flow of 850,000 gpd requires a dose of 25 mg/L chlorine. If sodium hypochlorite is 15 percent available chlorine, how many pounds per day are needed?

Answer: 25 mg/L × 0.85 mgd × 8.34/0.15 = 1,181 lb/d sodium hypochlorite.

Normally, these compounds are fed as solutions, and so it is necessary to determine the gallons-per-day feed necessary to deliver the required pounds per day. If the specific gravity is known, multiply it by 8.34 lb/gal and use that number as a divisor of the pounds per day to calculate gallons per day. If the specific gravity is unknown, assume it to be 1, and divide the pounds per day by 8.34 to calculate gallons per day.

Breakpoint Chlorination

Many water systems attempt to produce "free available" chlorine residual in their plant effluent. This process of adding chlorine to water until the chlorine demand has been satisfied is known as *breakpoint chlorination.*

Chlorine in this form has its highest disinfectant ability. Further addition of chlorine will produce a chlorine residual that is directly proportional to the amount of chlorine added beyond the breakpoint. Public water supplies are normally chlorinated past the breakpoint unless they practice chloramination.

Figure 8-3 is a graph showing the breakpoint chlorination curve. As an example, assume that the water being treated contains some manganese, iron, nitrite, organic matter, and ammonia. When a small amount of chlorine is added, it reacts with (oxidizes) the manganese, iron, and nitrite. And that is all that happens—no disinfection takes place and no chlorine residual forms (point 1 to point 2 on the graph). If a little more chlorine is added, enough to react with the organics and the ammonia, chloro-organics and chloramines will form. The chloramines produce combined chlorine residual—a residual that has combined with other components and has lost some of its disinfecting power. These components may cause tastes and odors (from points 1 to 3).

If just a little more chlorine is added, the chloramines are destroyed (points 3 to 4) and breakpoint, the point at which a little more chlorine will begin to produce a free available chlorine residual (beyond point 4), has been reached. This chlorine is free in the sense that it has not reacted with anything but it is free to do so and it will react if anything gets into the water that needs to be disinfected or oxidized further.

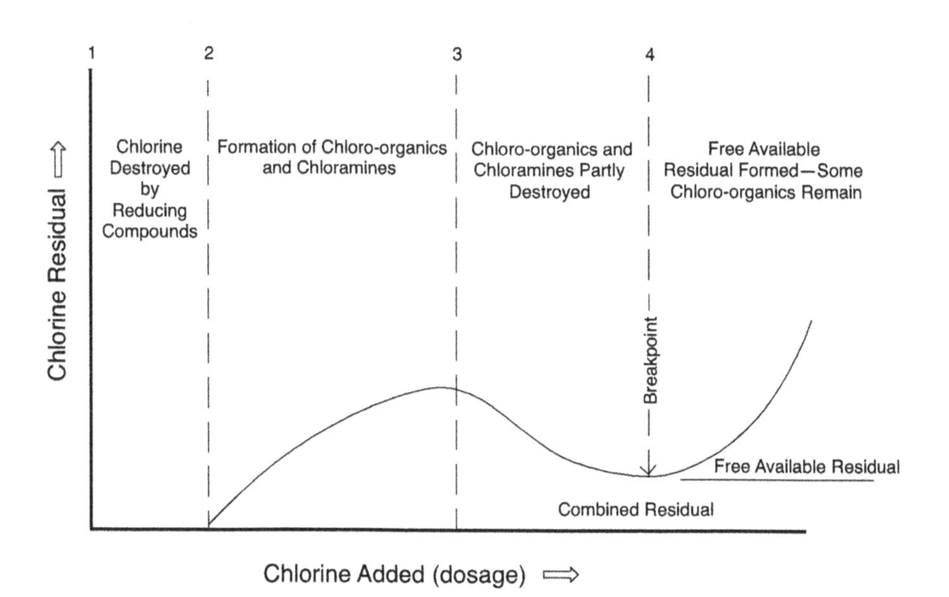

Figure 8-3 Breakpoint chlorination curve

CHLORINE FEED EQUIPMENT

Chlorine can be fed during pretreatment or post-treatment and often is fed during both to accomplish several tasks (Figure 8-4). As discussed in chapter 5, coagulation is usually enhanced in the presence of a strong oxidant, and chlorine is routinely used for this purpose in the pretreatment scheme. Use of pretreatment chlorine is sometimes minimized if disinfection by-product (DBP) formation is an issue.

Chlorine gas is fed using a pressure-feeder or vacuum-feeder system. Most often, the vacuum type is used. This equipment operates by sending a vacuum signal to the gas container. A mechanism draws the gas to a stream of water and produces a solution, which is fed at the application point(s). These units are safer than pressure feeders and generally provide more even dosage control.

CHLORINE DIOXIDE

Chlorine dioxide is a greenish yellow gas at room temperature and is odorous like chlorine. Because it is unstable at high concentrations, it is never shipped in bulk but rather generated on-site. Gaseous chlorine dioxide is generated from solutions of sodium chlorite that have been reacted with chlorine solutions at low pHs. In solution, the chlorine dioxide remains in its molecular form, ClO_2. The gas formation is shown in the following two reactions:

$$\text{Using chlorine: } 2NaClO_2 + Cl_2(g) \rightarrow 2ClO_2(g) + 2Na^+ + 2Cl^-$$

$$\text{Using hypochlorous acid: } 2NaClO_2 + HOCl \rightarrow 2ClO_2(g) + 2Na^+ + Cl^- + OH^-$$

Excess chlorine is often used in the generation step to drive the reaction to completion and to avoid the presence of unreacted chlorite. Chlorine dioxide can be dissolved in water and is stable in the absence of light and elevated temperatures. At higher pH and at higher temperatures, chlorine dioxide can revert to chlorite and chlorate, which are undesirable in drinking water. Chlorine dioxide was developed to aid in taste-and-odor control but is also used to destroy iron and manganese. Its use as a primary disinfectant for water supplies has gained popularity because it produces less DBPs than chlorine and is effective against *Cryptosporidium*.

Taste-and-odor problems may occur at residuals greater than 0.5 mg/L. Because it introduces chlorite to the water, which can cause certain types of anemia, its use is limited to a USEPA maximum residual disinfectant level (MRDL) of 0.8 mg/L.

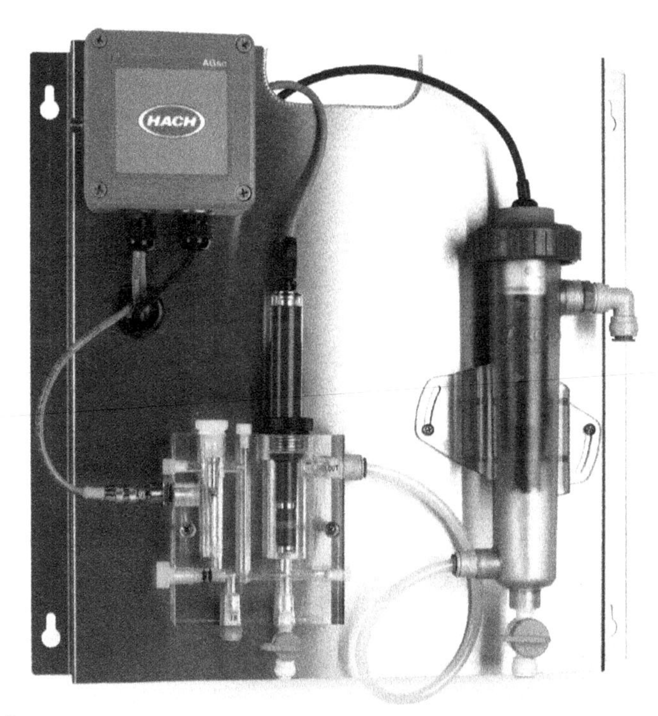

Source: Hach Company.

Figure 8-4 Chlorine residual analyzer

CHLORAMINATION

The practice of chloramination, while not a new technology for disinfection (see Table 8-2), has become popular for water systems that are trying to comply with more stringent DBP regulations. Originally developed to reduce treatment costs and to reduce tastes and odors caused by chlorine, chloramination is now recognized for its ability to prevent DBP formation and to extend chlorine residuals throughout a system. Plants that chloraminate usually add gaseous (or liquid) ammonia at post-treatment, after chlorinated water has traveled through a tank or clearwell for primary disinfection ($C \times T$) purposes. This is referred to as *post-ammoniation*. Many utilities use a chlorine:ammonia ratio of 4.5:1 to 5:1 to control nitrification in the distribution system.

The ammonia is supplied either as a liquefied, compressed gas in cylinders, horizontal containers, trucks, and railcars, as liquid aqua-ammonia, or as granular ammonium sulfate or ammonium chloride. Ammonia compressed gas is considered 100 percent strength, while liquid or aqua ammonia is usually 26° Baumé, which is 29.4 percent ammonia by weight. Ammonium sulfate granules are usually about 21 percent ammonia-N by weight.

Table 8-2 Utilities with years of experience using chloramination

City	Approximate Start of Chloramination
Denver	1914
Portland	1924
St. Louis	1934
Boston	1944
Indianapolis	1954
Minneapolis	1954
Dallas	1959
Kansas City	1964
Milwaukee	1964
Jefferson Parish (LA)	1964
Philadelphia	1969
Houston	1982
Miami	1982
Orleans Parish (LA)	1982
San Diego	1982

Source: Trussel and Kreft 1984.

Ammonia cylinders are different than chlorine cylinders in that they tend to be wider and are not equipped with a pressure-relief system. A 1-ton container of ammonia has two fusible plugs and holds 800 lb of material. Because ammonia is fed at about 25 to 33 percent of the amount of chlorine used, cylinders rather than larger units (trucks or railcars) are most often found at water treatment plants.

Ammonia is a dangerous gas, and safety precautions must be used at all times in areas where it is stored and used. It is readily detectable at concentrations of 20 to 50 ppm and induces general discomfort at concentrations of 150 to 200 ppm. It may be fatal at concentrations as low as 2,000 ppm.

Aqua ammonia can be fed from bulk tanks or barrels. Chemical metering pumps are used to deliver the chemical and pace the addition to match the chlorine dosage. The pumps must be equipped with systems to prevent feed malfunctions due to chemical off-gassing. Positioning pumps so that there is a positive suction head usually avoids this problem.

Granular ammonium sulfate can either be fed with dry chemical feeders or mixed with water for solution feed. Depending on the pH of the solution, the solution day tank may need to be sealed to avoid exposure to ammonia gas.

Because the feed of ammonia, as a ratio of the amount of chlorine used, is so important, the ammonia feeder is usually tied directly to the chlorine residual and water flow. Thus, compound-loop control systems are generally used, although some water treatment plants use manual control of ammonia feed.

The disadvantages of chloramine usage are

- It is a weak oxidant and disinfectant, especially against certain viruses.
- It may cause taste-and-odor problems.
- Kidney dialysis patients cannot tolerate it and so must be notified of its use.
- It can lead to nitrification problems in the distribution system.

Nitrification is an operational concern, especially in warmer waters. Nitrification is the process by which ammonia present in the water system is oxidized sequentially to form nitrites and then nitrates. If operators do not maintain the proper chlorine:ammonia ratio, the excess ammonia can convert to nitrites, and blooms of heterotrophic bacteria can abound. Most studies indicate an optimum chlorine:ammonia ratio is between 4.5:1 and 5:1 by weight.

Nitrifying bacteria grow as water temperature increases, and these slimes find refuge in low-flow areas and in tanks with low turnover rates. This causes a loss of residual and creates a higher chlorine demand. Systems then must resort to flushing and tank turnover programs to remediate the problem. Some systems resort to free chlorination once per year to burn out slimes that may accumulate. When this is done, the public must be notified of the change.

The City of St. Louis Water Department uses this strategy to keep the nitrification problem in check. When water temperature increases above 15°C (59°F), the utility will

- Maintain a minimum 4:1 chlorine:ammonia ratio by weight
- Maintain a finished water chloramine residual concentration equal to or greater than 2.5 mg/L
- Elevate finished water pH to equal to or greater than 9.5 (this is a lime-softening plant)
- Increase turnover in the distribution system and reservoir system to shorten residence time
- Monitor results of heterotrophic plate count bacteria counts using R_2A agar

OZONE

Ozone was first used as a disinfectant in the Netherlands in 1893. A few years later, a permanent installation for a continuous ozonator was built in Nice, France. Until the outbreak of *Cryptosporidium* in 1993 in Milwaukee, Wisconsin, only a handful of US water treatment plants used ozone. Since 1993, many utilities have resorted to its use.

Ozone (O_3) must be generated on-site, because it is an unstable gas (Figure 8-5). The formula for the ozone reaction is

$$O_2 + \text{energy} \rightarrow O + O$$

$$O + O_2 \rightarrow O_3$$

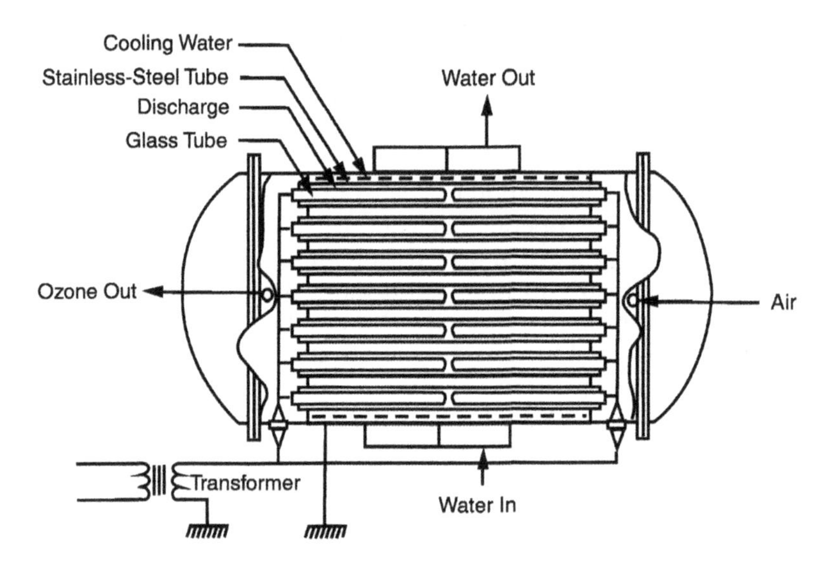

Figure 8-5 Schematic of an ozone generator

The elemental, or nascent, oxygen produced in the first reaction is provided by an electrical discharge with large amounts of voltage. The air or oxygen that is used must be dry and refrigerated, and it must be particle-free. The yield of ozone can be up to 14 percent by volume.

Ozone is very unstable in water, and maintaining a residual is difficult. High-alkalinity waters allow for a more stable ozone residual, which is very reactive with natural organic material (NOM) in the source water. When the residual reacts with the organic contaminants, it oxidizes them to a lower-molecular-weight species, including a variety of organic acids. Because these acids may contribute to biofouling of the distribution system, they are usually controlled using biologically active filtration. Ozone may also react with waters containing bromide and thus produce bromoform and other DBPs.

It is difficult to understand or study how ozone attacks and kills pathogens because it is difficult to measure low concentrations of dissolved ozone.

Ozone off-gas collection systems are necessary because ozone is toxic. It must be kept within Occupational Safety and Health Administration limits in the plant and surrounding areas and may be regulated as a discharge gas in some cases. The dissolution of ozone in water produces such a corrosive product that stainless-steel tanks and fluorocarbon gaskets are a must; basins used are made of concrete and caulked with an inert material.

ULTRAVIOLET LIGHT

Unlike ozone, the inactivation of microorganisms by ultraviolet (UV) light is well understood. The reaction is purely physical and therefore not significantly affected by pH. UV light is generated by mercury or antimony vapor lamps and is commonly used to treat wastewater effluent. UV disinfection is often considered for disinfection due to its effectiveness in inactivation of *Cryptosporidium* and because it does not produce DBPs. However, because it does not produce a lasting disinfectant residual, other technologies for residual maintenance (like chlorination or chloramination) are necessary.

UV disinfection is ideally suited to groundwater treatment. The process requires a stable, low-turbidity water. UV does not provide a protective residual so it is useful in groundwater systems that have a variance that allows them to forego disinfection. UV may then be used as an extra level of protection where residual is not required. Small water systems are also a good fit for UV because most small systems require only a basic level of operator skill to operate.

Process Description

UV light effectively destroys bacteria and viruses. However, how well the UV system works depends on the energy dose that the organism absorbs. An effective dose is measured as a product of the lamp's intensity, including radiation concentration, proper wavelength, exposure time, water quality, flow rate, and the microorganism's type and source, as well as its distance from the light source.

Performance

The LT2ESWTR lists log credit for UV disinfection of *Cryptosporidium, Giardia lamblia,* and *virus*. These values are listed in Table 8-3. Additionally, UV light disinfection does not form any significant disinfection by-products.

UV effectiveness is relatively insensitive to temperature and pH. UV application does not convert nitrates to nitrites, or bromide to bromines or bromates. Studies show that UV-treated drinking water inhibits bacterial growth and replication in the distribution system; however, conditions within distribution systems, such as leaks, still require additional residual disinfection (e.g., free chlorine).

Table 8-3 UV Dose (mJ/cm²) Table with Inactivation Credit

Log Credit	*Cryptosporidium* Dose	*Giardia lamblia* Dose	Virus Dose
0.5	1.6	1.5	39
1.0	2.5	2.1	58
1.5	3.9	3.0	79
2.0	5.8	5.2	100
2.5	8.5	7.7	121
3.0	12	11	143
3.5	15	15	163
4.0	22	22	186

Design and Operation

At a minimum, drinking water systems should install two UV units, both of which should be capable of processing the amount of water the system was designed to handle. Having two units in place ensures continuous disinfection when one unit is being serviced. Two units also can ensure operation during low-flow demand periods. Modular units designed for small drinking water systems are easy to install and operate (two plumbing connections per unit and one electrical hook-up). They should be equipped with automatic cleaners and remote alarm systems. For systems in isolated areas, operators should maintain and store a set of spare parts on-site, and consider a telemetry system for monitoring treatment.

$C \times T$ Values

The adequacy of disinfection is determined by the product of the final residual concentration (C) of the disinfectant and the contact time (T) for that disinfectant. The contact time is evaluated as that which is exceeded by 90 percent of the water, i.e., 10 percent of the water has less contact time than the T value. These T values are commonly determined by performing tracer studies or by computations based on the ratio of the length and width of the reaction chambers (clearwells or contactors), including any baffling that may be present. Baffling of clearwells, which involves adding separation walls that force snakelike flow patterns, provides greater length-to-width ratios and therefore greater contact times for the disinfectant.

The USEPA has published tables that show the $C \times T$ values needed for basic disinfectants such as chlorine, chloramine, ozone, and chlorine dioxide. $C \times T$ values will vary with temperature, pH, and concentration. Surface water treatment plants must provide the minimum $C \times T$ value in order to be in compliance with the SWTR. The $C \times T$ calculation, which is performed every 24 hours, is derived at the highest hourly flow rate and lowest contactor volume for that period. Refer to appendix B for an example $C \times T$ calculation.

DISINFECTION BY-PRODUCTS

There are many DBPs, the chief of which are total trihalomethanes (TTHMs) and haloacetic acids (HAAs). These are formed when chlorine is added to water that contains the necessary types and amounts of NOM.

Chapter 1 discusses the regulatory concerns for DBPs, but operators should understand the kinetics of DBP formation and know which mechanisms are under their control. In general, the following may be considered true for DBP formation:

- DBP formation is affected by contact time between the disinfectant and the NOM in water. Therefore, minimizing that contact time, either by adding disinfectant later in the treatment process or by "turning over" the water in distribution storage tanks more frequently, may decrease DBP formation.

- The *water age*–the amount of time that water resides in the distribution system or storage tanks before being consumed—is of critical importance. Longer water age usually leads to increased formation of DBPs.
  - Water age can be estimated by simply performing *tracer studies* of the distribution system. Operators choose representative sample locations in the system, and begin sampling them each day for a tracer such as fluoride. The fluoride feed at the plant is turned off (with permission of the regulatory authority), and the time it takes for the fluoride at each sample location to reach background levels will give a good snapshot of relative water age at these locations. Operators can then make improvements by changing valve positions, booster settings, and so forth, which may decrease this age.
  - Another way to estimate water age is to use a *calibrated* system hydraulic model. Many hydraulic modeling computer programs can generate calculated water age values for locations throughout the distribution system.
- Increasing disinfectant dose and residual increases DBP formation. **Caution: Operators must avoid the risk of pathogen passage that can accompany disinfectant losses. A balance must be achieved to simultaneously remove microbiological contaminants while controlling DBP formation.**
- As temperature increases, DBP formation may increase. Warmer water favors the formation of DBPs, as does summer heat, which bakes the system storage tanks.
- The water pH can control the type of formation—lower pH favors HAA formation, while higher pH favors TTHM formation.
- The nature and concentration of the precursor material will affect the types and amounts of DBP formation. Bromide in the water, or as a contaminant in the chemical feed system, will influence formation of brominated DBP species.

BIBLIOGRAPHY

AWWA. 2006. M20—*Manual of Chlorination and Chloramination Practices and Principles*, 2nd Ed. Denver, Colo.: American Water Works Association

Bolton, J., and C.A. Cotton. 2008. *The Ultraviolet Disinfection Handbook*. Denver, Colo.: American Water Works Association.

Edzwald, J.K., ed. 2011. *Water Quality and Treatment*, 6th ed. New York: American Water Works Association and McGraw-Hill.

Lauer, W.C., M. Barsotti, and D.K. Hardy. 2009. *Chemical Feed Field Guide for Treatment Plant Operators*. Denver, Colo.: American Water Works Association.

Trussell, R. and P. Kreft. 1984. AWWA Seminar Proceedings: Chloramination for THM Control. Denver, Colo.: American Water Works Association.
USEPA. 2007. Simultaneous Compliance Guidance Manual for the Long Term 2 and Stage 2 DBP Rules, USEPA, Office of Water (4601) EPA 815-R-07-017
USEPA. 2006. Ultraviolet Disinfection Guidance Manual for the Final Long Term 2 Enhanced Surface Water Treatment Rule, USEPA 815-R-06-007
Visintainer, D. 2001. City of Saint Louis Water Commissioner. Internal communication to author. October.

Chapter 9

Softening

The two traditional ways to soften hard water are the lime–soda ash process and the ion-exchange process. Membranes, including nanofiltration (NF) and reverse osmosis (RO), can also be used to soften water, and they have gained an increasing use in this field. Membranes are discussed in detail in chapter 11. Electrodialysis (ED) and electrodialysis reversal (EDR) are other processes that can be used to soften water. They are used mostly for demineralization of high total dissolved solids (TDS) water and a description of the process is included in chapter 10.

WATER HARDNESS

As outlined in *Hoover's Water Supply and Treatment* published by the National Lime Association, the hardness of most water supplies is due to four compounds that are held in solution. These compounds are

- Calcium bicarbonate (solution of limestone, $CaCO_3$, in waters containing carbon dioxide, CO_2)
- Magnesium bicarbonate (solution of magnesite, $MgCO_3$, in waters containing carbon dioxide, CO_2)
- Calcium sulfate (in the form, $CaSO_4 \cdot 2H_2O$, known as gypsum)
- Magnesium sulfate (in the form, $MgSO_4 \cdot 7H_2O$, known as Epsom salts)

If one or a combination of these compounds is present in a water supply, the water has a degree of hardness. Softening of water requires the reduction or removal of calcium and magnesium from the water—the more that is removed, the softer the water will be.

Calcium and magnesium also form compounds with chlorides and nitrates, which are sometimes present in water supplies and would also contribute to overall hardness. Calcium and magnesium bicarbonates, formerly designated as temporary hardness because they could be partially precipitated by boiling the water to drive off carbon dioxide, are now termed *carbonate hardness*. Calcium and magnesium sulfates, chlorides, and nitrates, formerly designated as permanent hardness, are now called *noncarbonate hardness*.

Iron, manganese, and other divalent cationic ions also cause hardness. However, because they are usually present in small quantities, it is customary not to consider them in connection with hardness. However, the presence of iron and manganese is very objectionable (see chapter 10) because they stain materials with which they come in contact, such as porcelain bathroom fixtures, and are especially objectionable when water is used for laundry purposes. Iron and manganese can be removed separately using oxidation treatment processes, but the softening processes also remove iron and manganese as well as the other hardness compounds.

Total Hardness

The total hardness of water is the sum of the bivalent (two positive charges) metallic cations, primarily calcium and magnesium. Chemical analyses are performed to determine this value. Lab analysts normally determine total hardness and calcium hardness and then subtract the two to get the magnesium hardness. The value is expressed in terms of milligrams per liter as calcium carbonate. For example, if calcium hardness is determined to be 90 mg/L as $CaCO_3$ and the total hardness is 125 mg/L as $CaCO_3$, the magnesium hardness is 35 mg/L magnesium (125 – 90) as $CaCO_3$. If the values for calcium and magnesium are expressed as the cations, a multiplier can be used to make the conversion so the value can be expressed as $CaCO_3$. These multipliers are derived from the ratio of the atomic weight of $CaCO_3$, which is 100, and the atomic weights of the metals. The atomic weight of calcium is about 40 and magnesium is about 24.3; the multiplier is 2.5 for calcium and 4.1 for magnesium.

Sample Calculation

The lab gives the operator a calcium value of 42 mg/L as calcium and a magnesium value of 12.2 mg/L as magnesium. The operator wishes to express the total hardness as $CaCO_3$.

$$42 \text{ mg/L calcium} \times 2.5 = 105 \text{ mg/L as } CaCO_3$$

$$12.2 \text{ mg/L magnesium} \times 4.1 = 50 \text{ mg/L as } CaCO_3$$

$$\text{Total hardness, as } CaCO_3 = 105 + 50 = 155 \text{ mg/L}$$

Alkalinity

The alkalinity of water is a measure of its carbonate hardness and is determined in the laboratory by titration analysis (see chapter 12). Carbonate hardness is hardness that is chemically equivalent to the alkalinity where most of the alkalinity in natural waters is caused by carbonate and bicarbonate ions. Noncarbonate hardness is then equal to the hardness minus the alkalinity and represents the calcium and magnesium that are bound up with sulfates, chlorides, and nitrates. In the previous hardness calculation, an alkalinity measurement of 100 mg/L as $CaCO_3$ would yield a noncarbonate hardness of 55 mg/L (155 – 100).

Sometimes, the alkalinity is greater than the hardness. In those instances, the hardness is equivalent to the carbonate hardness and the noncarbonate hardness is zero.

The relationship between hardness and alkalinity must be understood in order to perform the calculation common to the precipitative softening process. For operational purposes, think of alkalinity as existing in one or more of three forms: hydroxide (OH^-), bicarbonate (HCO_3^-), and carbonate (CO_3^{-2}). If any of these three are linked to calcium or magnesium, there is carbonate hardness. The lab test (titration with a standard acid solution to a pH end point of 4.5) that determines total alkalinity (methyl orange test) yields a value, which is labeled as *T*. Another test, the phenolphthalein test (titration to the pH endpoint of 8.3), yields a value of *P*. The *P* alkalinity is related to the *T* alkalinity in one of five ways, as shown in Table 9-1. The *P* alkalinity can be 0, less than ½ *T*, equal to ½ *T*, greater than ½ *T*, or equal to *T*. The table enables the operator to determine the relationship that the three types of alkalinity have simply by using the *P* and *T* relationship. For example, if the lab shows a *T* alkalinity of 48 mg/L and a *P* alkalinity of 31 mg/L, the relationship of *P* to *T* is that *P* is greater than ½ *T*. From the table, it is noted that the hydroxides must be $2P - T$, or 14; the carbonates must be $2(T - P)$, or 34; and the bicarbonate is 0.

PRECIPITATIVE SOFTENING PROCESS

The precipitative softening process uses chemical reactions that render dissolved calcium and magnesium insoluble. Chemicals such as lime, caustic soda, and soda ash are added to the water in sufficient quantities to create these insoluble compounds (Figure 9-1). As insoluble compounds, or precipitates, the calcium and magnesium can be settled out and/or filtered, which results in a softer water. The word *precipitate* is used to designate a newly created chemical that has been formed by compounds that were previously dissolved. A precipitate is designated using a symbol ↓. When this symbol follows a chemical compound, it means that the compound is insoluble and will settle out or be filtered under normal operating conditions.

Table 9-1 Relationship of *P* and *T* alkalinities

	OH^-	CO_3^{-2}	HCO_3^-
$P = 0$	0	0	T
$P < ½T$	0	$2P$	$T - 2P$
$P = ½T$	0	T	0
$P > ½T$	$2P - T$	$2(T - P)$	0
$P = T$	T	0	0

Note: *P* = phenolphthalein alkalinity, *T* = total or methyl orange alkalinity.

Source: US Filter.

Figure 9-1 Lime storage silo

Chemistry of Softening With Lime, Caustic Soda, and Soda Ash

The reactions that take place when lime is added to hard water containing both calcium and magnesium are

$$Ca(HCO_3)_2 + Ca(OH)_2 \rightarrow 2CaCO_{3\downarrow} + 2H_2O \qquad \text{(Eq. 9-1)}$$

$$Mg(HCO_3)_2 + Ca(OH)_2 \rightarrow MgCO_3 + CaCO_{3\downarrow} + 2H_2O \qquad \text{(Eq. 9-2)}$$

$$MgCO_3 + Ca(OH)_2 \rightarrow Mg(OH)_{2\downarrow} + CaCO_3 \qquad \text{(Eq. 9-3)}$$

$$MgSO_4 + Ca(OH)_2 \rightarrow Mg(OH)_{2\downarrow} + CaSO_4 \qquad \text{(Eq. 9-4)}$$

$$MgCl_2 + Ca(OH)_2 \rightarrow Mg(OH)_{2\downarrow} + CaCl_2 \qquad \text{(Eq. 9-5)}$$

$$Mg(NO_3)_2 + Ca(OH)_2 \rightarrow Mg(OH)_{2\downarrow} + Ca(NO_3)_2 \qquad \text{(Eq. 9-6)}$$

The reactions involving the use of soda ash are

$$CaSO_4 + Na_2CO_3 \rightarrow CaCO_{3\downarrow} + Na_2SO_4 \quad \textbf{(Eq. 9-7)}$$

$$CaCl_2 + Na_2CO_3 \rightarrow CaCO_{3\downarrow} + 2NaCl \quad \textbf{(Eq. 9-8)}$$

$$Ca(NO_3)_2 + Na_2CO_3 \rightarrow CaCO_{3\downarrow} + 2NaNO_3 \quad \textbf{(Eq. 9-9)}$$

Note that $MgCO_3$ is formed in Eq. 9-2, and additional lime is added (Eq. 9-3), to convert this $MgCO_3$ to $Mg(OH)_2$. In Eq. 9-4, 9-5, and 9-6, $Mg(OH)_2$ is formed, but the calcium from the lime forms soluble compounds (i.e., $CaSO_4$, $CaCl_2$, and $Ca(NO_3)_2$). Therefore, soda ash must be added (Eq. 9-7, 9-8, and 9-9) to convert these soluble compounds to $CaCO_3$.

The typical water softening plant will analyze the source water for the following four characteristics:

- CO_2 in solution, if present
- Alkalinity (total and phenolphthalein)
- Total calcium and magnesium (total hardness)
- Noncarbonate hardness

From these data, an estimate of the lime and soda ash requirements of the softening process can be produced:

- For aqueous CO_2, lime must be added.
- For calcium carbonate hardness, lime must be added.
- For magnesium carbonate hardness, twice as much lime must be added.
- For calcium noncarbonate hardness, soda ash must be added.
- For magnesium noncarbonate hardness, lime and soda ash must be added.

The balanced chemical reactions can be used to approximate the lime and soda ash required to soften a water completely (i.e., to a minimum achievable total hardness of 40 to 60 mg/L, as $CaCO_3$) (Figure 9-2). If a higher total hardness is desired in the finished water, less lime or soda ash is added than is required to soften the water completely.

Caustic soda may be used to remove both carbonate and noncarbonate hardness from water. However, when caustic soda is added to water containing soluble calcium and magnesium salts, it precipitates the magnesium as insoluble magnesium hydroxide and absorbs carbonic acid, resulting in the formation of an equivalent amount of sodium carbonate (soda ash). If there is enough noncarbonate hardness in the water being treated to react with the sodium carbonate formed, caustic soda may be used as a softening agent instead of lime and soda ash. As a general rule, caustic soda can replace all the soda ash but only part of the lime. With the higher cost associated with caustic soda, there would be little if any savings involved. Laboratory tests under these conditions indicate that the use of caustic soda may leave more calcium carbonate in the settled water than soda ash treatment.

Regardless of whether lime, soda ash, or caustic soda is used to chemically soften hard water, operators should remember that the chemical

Source: Sangre de Cristo Water Division.

Figure 9-2 Proper storage of soda ash. Bags are carefully stacked within floor loading limits.

reactions used to approximate dosages do not account for the purity of the softening chemical being added. The better commercial grades of quicklime are approximately 90 to 94 percent CaO, and high-quality calcium hydrated lime is approximately 68 to 73 percent CaO. The best grades of soda ash are approximately 99 to 100 percent Na_2CO_3, and caustic soda is typically 98 to 99 percent NaOH. Therefore, quantities of pure chemicals approximated from calculations using the reactions should be divided by the appropriate percent purity and multiplied by 100 to determine the quantity of commercial-grade chemical that must be added.

Determining Proper Quantity of Lime

If the free CO_2 of the water is known and the magnesium and alkalinity, expressed as $CaCO_3$, are known, the lime requirement can be calculated using the following equation:

$$[10.6\ (CO_2) + 4.7\ (\text{alkalinity}) + 4.7\ (Mg^{+2})]/\text{purity of lime, in percent} \quad \textbf{(Eq. 9-10)}$$

Use of Eq. 9-10 will give the lime requirement in pounds per million gallons. The constants (10.6 and 4.7) are used to convert milligrams per liter to pounds per million gallons. It is usually necessary to carry an additional or excess amount of lime in the water after treatment, and so operators add 10 to 50 mg/L to the calculated amount. Determining the lime feed (Figure 9-3) requirement in this manner is one of many process control strategies that can be used. For example, some plants attempt to maintain the pH in the softening zone at a specific level. With this approach, the finished water pH

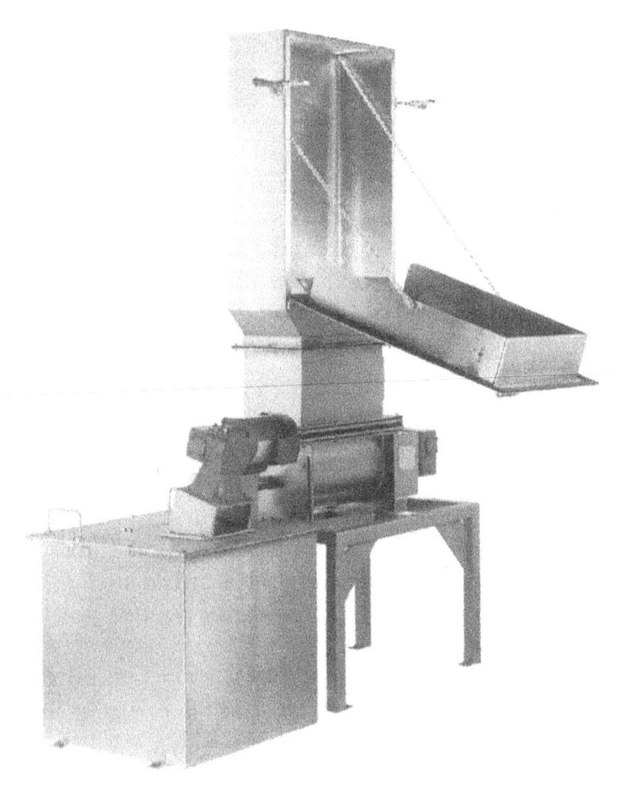

Source: Acrison Inc.

Figure 9-3 Lime feeder

varies more than for a plant that controls its processes to maintain a consistent finished water pH by varying the pH in the softening zone.

An example of how one utility establishes its lime dosage is found in the operational techniques at the city of St. Louis, Missouri. St. Louis uses a raw river water source, which is highly variable, and consequently bases its lime feed rates on three criteria:

1. Maintenance of softening zone pH in the range of 9.8 to 10.2
2. The ratio of phenolphthalein alkalinity (*P*) to total alkalinity (*T*) in the softening zone should not exceed 0.5
3. Total river hardness

Generally, to increase the pH 0.3 units as it travels through the treatment process, St. Louis must add an additional 0.3 gpg (5.14 mg/L) of lime. If the *P*/*T* ratio is higher than 0.50, the lime dosage is decreased to lower this ratio. A review of historical data for St. Louis indicates that the optimal *P*/*T* ratio is actually 0.49 in the summer and 0.45 in the winter at one treatment facility. Changes in total river hardness are also used to anticipate changes in pH and to adjust lime dosages. For every 20-mg/L rise in river hardness, a 0.3-gpg (5.14-mg/L) increase in lime dosage is generally required. It is important to note that such criteria may vary somewhat from plant to plant and between different raw water sources.

Recarbonation

After water is softened, it is supersaturated with lime and may need to be recarbonated. Recarbonation is the process of adding CO_2 to the water, usually before filtration, to convert hydroxide and carbonate alkalinity to bicarbonate alkalinity. During the process, the pH will drop to desirable levels. The amount of CO_2 (in pounds per million gallons) necessary for recarbonation can be calculated using the formula:

$$3.67\ (OH^- \text{ change} + CO_3^= \text{ change}) \qquad \textbf{(Eq. 9-11)}$$

where OH^- change and $CO_3^=$ change are the amount, in milligrams per liter, these two components decrease from before recarbonation to after recarbonation. Table 9-1 can be used to calculate the drop. For example, if settled water with a T alkalinity of 70 mg/L and a P alkalinity of 40 mg/L is treated with CO_2 and the operator desires a recarbonated T alkalinity of 60 mg/L and a P alkalinity of 28 mg/L, how many pounds per million gallons of CO_2 are needed? Using Table 9-1, before recarbonation, $P > ½T$ ($T = 70$ and $P = 40$) and, after recarbonation, $P < ½T$ ($T = 60$ and $P = 28$). Therefore, the hydroxides must have been $2P - T$ (80 – 70), or 10 mg/L before recarbonation. After recarbonation, the hydroxides are 0 according to the table. So the total drop in hydroxides is 10 – 0, or 10 mg/L. The drop in carbonates is also found using the table. Before recarbonation, the carbonates must have been $2(T - P)$, or 2(70 – 40), 60 mg/L. After recarbonation, the carbonates drop to $2P$, or 56 mg/L, which is a change of 4 mg/L from before to after.

Using Eq. 9-10, the amount of carbon dioxide needed to make this change can be calculated:

$$\text{lb/mil gal } CO_2 = 3.67\ (10 + 4) = 51.38 \text{ lb/mil gal}$$

Amounts of carbon dioxide used will vary from plant to plant and will be affected by water chemistry and the desired finished water pH. In general, low-magnesium waters will not require excess lime for softening and so will not require as much recarbonation as high-magnesium waters. Recarbonation can take place in a single stage or in a two-stage process. A single stage is used in low-magnesium waters. Carbon dioxide is fed after sedimentation; operators will normally use enough carbon dioxide to reach a pH of 8.5 to 9.0 just before filtration. This will dissolve most of the calcium carbonate crystals—the rest are removed through filtration. In the two-stage process (usually used for high-magnesium waters), the raw water pH is elevated to 11 or greater with lime. After that stage, carbon dioxide is added to lower the pH to 10.0–10.6, where the optimum value for calcium carbonate precipitation is reached. After soda ash treatment, more carbon dioxide is added (second stage) to bring the pH to 8.4–8.6. Although operating costs are lower, it is initially expensive because two sets of basins are needed. For this reason, many high-magnesium waters are recarbonated with only one stage.

Carbon dioxide is furnished in pure form as a pressurized liquid by tank trucks, which transfer it to storage tanks. In the past, carbon dioxide was generated on-site by burning fuel; this is still done at a few plants. The liquid is vaporized and fed as a gas through a rotameter or sometimes through a chlorinator-type feeder. Operators should use caution in areas where carbon dioxide is stored. Although not a poisonous gas, carbon dioxide is colorless and odorless and heavier than air. Without warning, carbon dioxide can displace oxygen, and workers can suffocate in minutes.

Plants that soften at relatively low pHs or plants that soften early in the treatment process and then lower the pH to acceptable levels later by adding coagulants, disinfectants, etc., may not need to practice recarbonation. Utilities that do not recarbonate should monitor their filters carefully for carbonate deposition.

Chemical Feed Equipment

In addition to the chemical feeders and pumps that are used in any conventional water treatment plant, a precipitative softening plant needs gravimetric or volumetric feeders of sufficient size to handle the relatively large quantities of chemicals it will use. A glance at Table 9-2 gives the operator a perspective of the amounts of softening chemicals typically used, as compared to other treatment chemicals. This table lists the chemicals used, in pounds per million gallons, in an average day at the city of St. Louis water treatment plants. Note the difference in quantities. Because more softening chemicals are normally used in a treatment plant than other chemicals, the feed equipment and associated storage facilities are generally quite large and may need more attention and maintenance.

Hardness Tests

The EDTA (ethylenediaminetetraacetic acid) titrimetric test is the test most often used at water treatment laboratories to measure hardness level. It can be performed using standard chemical laboratory methodologies. Alternatively,

Table 9-2 Relative amounts and costs of chemicals used at St. Louis Water Department

Chemical	Pounds Used per Million Gallons	Cost per Million Gallons
Ferric sulfate	260	$12.25
Chlorine	37	$ 5.56
Fluoride	29	$ 1.97
Lime	803	$23.30

Source: City of Saint Louis, Mo., Water Department.

where ease of use and portability are needed, prepackaged kits can be used. Both methods depend on a titration step. Titration is the process of determining the amount of a substance present in a solution by measuring the amount of a different solution of known strength that must be added to complete a chemical change. In the hardness test, a dye that turns the sample red at pH 10 is added to the water being tested. A solution of EDTA, at known strength, is added to the sample drop-wise until the sample color changes to blue, which is an indication that all calcium and magnesium in the sample has been changed (complexed). The amount of hardness is a multiple of the amount of EDTA used to achieve the change. The operator should refer to *Standard Methods for the Examination of Water and Wastewater* or follow the instructions that come with the kit for proper procedures.

Alkalinity Tests

Like the hardness test, the test for alkalinity can be performed with titration methods. Because the total alkalinity test determines the level of hydroxide, carbonate, and bicarbonate activity in the water, it is standard practice to use an acid of known strength as the titrant. A weak solution of sulfuric acid (0.02*N*) is added to the sample that has been previously dosed with a dye. As the color changes, an end point is reached. This occurs at about pH 4.5.

To determine *P* alkalinity, a different color indicator is used. Traditionally, *P* alkalinity is determined by titrating the solution with weak acid until the pH drops to 8.3. The color indication method can be used or a pH meter can also be used.

Stability Test

Water that is under- or oversaturated with $CaCO_3$ tends to cause corrosion or scaling problems in the distribution system. For this reason, many precipitative softening plants produce treated water with a chemistry that deposits a slight amount of $CaCO_3$ onto the system piping. It is believed to be better to slowly encrust the system rather than corrode the piping with water that is undersaturated.

A stability test can be performed to determine the corrosiveness or scale-forming tendency of water. Two stoppered sample bottles are marked A and B; bottle A is spiked with 0.5 g of $CaCO_3$. Both bottles are then completely filled with water (no air is left in them). The bottles are held at room temperature, and bottle A is occasionally inverted to disperse the $CaCO_3$. Alternatively, a magnetic stirring bar and device are used to keep the $CaCO_3$ in suspension. Over time (1 to 2 hours), the water sample with the $CaCO_3$ either picks up some alkalinity or does not. After the suspension settles, a portion of the settled water from bottle A is filtered and analyzed for alkalinity. A comparison can then be made to the alkalinity of the water in bottle B. If the alkalinity in bottle A is higher than that in bottle B, the water may be corrosive. If the alkalinity in bottle A is lower than that in bottle B, the water may be scale-forming.

Langelier Index

For an indication of a water's corrosivity or scale-forming properties, the saturation index, or Langelier index (LSI) should be used.

$$LSI = pH\ (measured) - pHs$$

- For LSI > 0, water is super saturated and tends to precipitate (deposit) a scale layer of $CaCO_3$.
- For LSI = 0, water is saturated (in equilibrium) with $CaCO_3$. A scale layer of $CaCO_3$ is neither precipitated nor dissolved.
- For LSI < 0, water is undersaturated and tends to dissolve (corrosion) solid $CaCO_3$.

In practice LSI between -0.5 and +0.5 rarely show corrosion or deposition characteristics.

The pHs (pH at saturation in the above equation) can be computed using a nomograph and analyses of total dissolved solids, pH, calcium, alkalinity, and temperature. This index predicts the final pH that a water should have to be in equilibrium. The reader should refer to ***Standard Methods for the Examination of Water and Wastewater*** for this test.

Residuals

The residue produced from the precipitative softening process consists mostly of calcium carbonate or a mixture of calcium carbonate and magnesium hydroxide. The sludge solids content is typically near 5 percent, although a range of 2 to 30 percent has been reported. Source waters that require coagulation with aluminum or iron salts will have residuals that contain those components as well.

The amount of dry solids produced during the softening process can be calculated using formulas that account for calcium and magnesium being precipitated and for the amounts of suspended solids removed and chemicals added. The dry weight of sludge solids, ΔS, in kilograms per day, is

$$\Delta S = 86.4Q(2.0Ca + 2.6\ Mg + 0.44\ Al + 1.9\ Fe + SS + A) \qquad \textbf{(Eq. 9-12)}$$

where:

Q = source water flow, m^3/sec
Ca = calcium hardness removed, mg/L as $CaCO_3$
Mg = magnesium hardness removed, mg/L as $CaCO_3$
Al = alum dose as 17.1 percent Al_2O_3
Fe = iron dose as Fe, mg/L
SS = suspended solids concentration in source water, mg/L
A = additional chemicals such as polymer, activated carbon, or clay, mg/L

Ion Exchange Softening Process

Calcium and magnesium can be removed from water by exchanging calcium and magnesium ions for others that do not impart hardness to the water. The ion-exchange process (Figure 9-4) involves the use of a resin that has been saturated with nonhardness-producing ions that readily participate in this exchange process. The process used in water treatment plants is reversible and therefore renewable for many cycles. The method of resin renewal is called *regeneration*. A bed of resin is said to be "exhausted" when it can no longer remove hardness. At that time, regeneration using salt water (brine) is initiated.

Exchange resins vary in their capacity to soften and be regenerated; it is common to describe them based on the number of bed volumes that can be exchanged before the resin must be regenerated. Also, the amount of calcium and magnesium in the source water dictates the amount of water that can be softened. Harder water exhausts a resin bed sooner than a softer water. If an operator knows the exchange capacity of the resin bed, the hardness of the water, and the flow rate of water through the bed, he or she can predict the length of a softening cycle. This prediction is an important part of the operational scheme in ion-exchange softening plants because it allows operators to schedule time for other tasks and even allows for softening units to operate unattended. More importantly perhaps, because most ion-exchange plants have more than one unit, it allows for staggered operation. In this way, all units are not regenerated at the same time.

Design of Resin Beds and Exchange Units

The hardware involved in the ion-exchange process (vessels, piping, valves, etc.) is important; however, it will not be covered in this discussion. Rather, an explanation of the materials used (resin, regenerant) is presented.

Ion-Exchange Media

The earliest known materials used to soften water by ion exchange were naturally occurring zeolites found in certain soils. Eventually, synthetic resins were developed that had faster exchange rates and higher capacities to soften. These natural and synthetic materials are rated according to their ability to exchange calcium and magnesium for sodium, an ion that does not impart hardness to the water. The rating, or exchange capacity, can be expressed in terms of kilograins (kgr) of hardness per cubic foot of resin. Typically, the modern synthetic resins have exchange capacities of 20 to 40 kgr/ft^3; natural zeolites were rated at 3 to 5 kgr/ft^3. Natural zeolites could process about 200,000 bed volumes before replacement was needed. Modern synthetic resins can tolerate about 10 times that number of bed volumes before replacement is indicated.

Source: US Filter.

Figure 9-4 Ion-exchange units

Multiple terms are used to express hardness, and operators need to know the relationship between these. Hardness expressed as milligrams per liter $CaCO_3$ can be converted to grains per gallon hardness using the following formulas:

$$1 \text{ grain/gal hardness} = 17.1 \text{ mg/L hardness, and}$$

$$1{,}000 \text{ grains} = 1 \text{ kilograin}$$

An ion-exchange unit in a water treatment plant that has a resin rated at 30 kgr/ft^3 can theoretically remove 513,000 mg of hardness (30 × 1,000 × 17.1) for each cubic foot of resin before it is exhausted. In practice, resin beds normally remove a high percentage of the theoretical exchange capacity because of several factors, including flow rates, bed fouling, ionic interference, and efficiency of regenerant distribution. Dead space in the vessel is also a factor because it is nearly impossible to evenly distribute flow through it. However, it is important to realize that until the bed is exhausted, it is considered able to produce water with zero hardness. Beds are usually operated at flow rates of 1 to 5 gpm/ft^2 (8 to 40 bed volumes per hour) and will remove 50 to 75 percent of the rated hardness-exchange capacity.

Flow Rate Through Bed

Each gallon of water that flows through the bed carries an amount of hardness with it. It is necessary to know the hardness and flow rate values because they will dictate the time of exhaustion. If a bed is loaded at 120 gpm (454 L/min) and the source water has a hardness of 14 gr/gal (240 mg/L), 1,680 gr (120 gpm × 14 gr/gal) of exchange capacity will be used each minute of operation. The operator can see that the bed will exhaust eventually and that the exhaustion time will be reached sooner if the flow rate is increased. Doubling the flow rate will theoretically exhaust the bed in half the time.

As stated, water leaving the ion-exchange vessel should have zero hardness. However, because water with zero hardness tends to corrode the system and the customers' plumbing, ion-exchange water plants must produce water that is not of zero hardness. The simplest way to do this is to bypass some of the source water around the softeners and then blend it with the zero-hardness water at a predetermined rate to achieve the desired level of finished water hardness. This predetermined rate is called the *percent bypass* and is calculated by dividing the desired finished water hardness by the source water hardness and converting that value to a percent. This number represents the approximate amount of water that should bypass the softeners. For example, if the source water hardness is 20 gpg (342 mg/L) and the desired finished water hardness is 8 gpg (137 mg/L), the percent bypass is calculated at 8/20, or 40 percent. If the plant flow is 1,000 gpm (2.2 ft^3/sec), the operator knows that 40 percent of 1,000, or 400 gpm (0.9 ft^3/sec), should bypass the softeners to get the desired results.

This bypass water will have the effect of lengthening the life of the softener. More finished water is produced during the cycle because not all of it had to be softened. The greater the desired hardness of finished water, the more efficient the plant becomes with respect to softening expense.

Regeneration

When the bed is exhausted, it must be regenerated. Because the resin has been stripped of sodium ions, these ions must be replaced. This is done by rinsing the bed with a salt solution (brine) at sufficient volume and concentration for a set time. Brine is concentrated sodium chloride (NaCl) solution, which is approximately a 25 percent concentration at 60°F (15.5°C). The specific gravity of this solution is about 1.192, which yields a working solution of about 2.479 lb of salt per gallon. The brine is usually diluted further before being applied to the softener in the regeneration cycle. As an example, if a softening unit contains 1,000 ft^3 of ion-exchange material and it has a hardness-removal capacity of 3,000 gr/ft^3, it will have a softening capacity of 3,000,000 gr. If it is assumed that 0.5 lb of salt is required per 1,000 gr of hardness removed, 3,000 × 0.5, or 1,500 lb of salt will be required to regenerate the unit. Each gallon of saturated brine contains 2.479 lb of salt; therefore, approximately 605 gal of brine solution will be required to regenerate the unit. To produce a 5 percent solution (1.03624 specific gravity), the 605 gallons of brine must be diluted to 3,480 gal.

An exhausted resin bed is first backwashed to remove any foreign particles and to restratify the bed. The diluted brine is then introduced through the bed slowly, usually for 20 to 30 minutes. The flow rate may be about 1 gpm/ft^2 (0.7 mm/sec). It is very important to provide adequate time for the regeneration effect to take place. The bed is then rinsed to remove unused brine before the unit is placed back into service.

Operating an Ion-Exchange Plant

Figure 9-5 is a schematic of an ion-exchange plant. Note that this plant treats well water that contains iron. Therefore, the plant is equipped with an aerator and iron-removal filters, which is a typical arrangement for ion-exchange plants treating water with iron problems (chapter 10). The iron must be removed before the water is passed through the exchangers; otherwise, the resin will be fouled and rendered useless.

Operators running this type of process must know the design and operating data of the plant. Designers of these plants use a much more complex set of variables, but the following examples represent a basic operational view of ion-exchange softening.

There are three ion-exchange softeners, each measure 6 ft (1.8 m) in diameter and have 7 ft (2.1 m) of media depth in them. The exchange capacity of the media is 15,000 gr/ft^3, and it takes 0.3 lb of salt per 1,000 gr of hardness removed to regenerate. The source water hardness is measured at 325 mg/L and the finished water hardness is 143 mg/L. One 500-gpm (1.1-ft^3/sec) high-service pump is operating. With this information, solve the following sample problems.

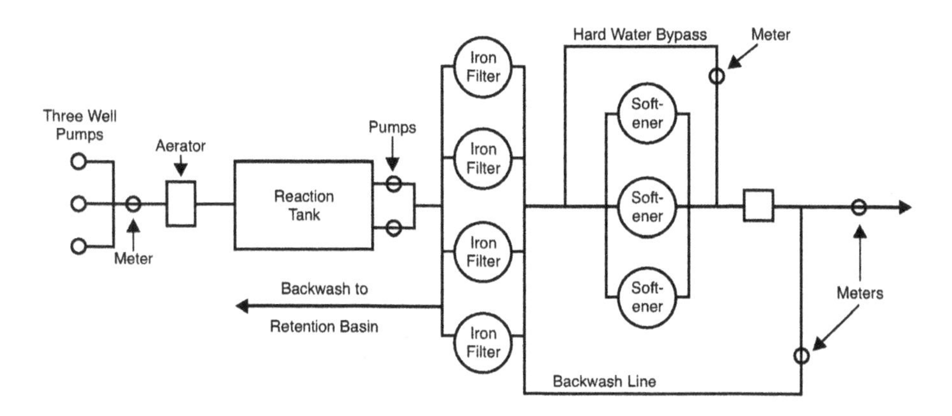

Figure 9-5 Schematic of the ion-exchange process used in the example problems

What is the total hardness-removal capacity of the plant?

The volume of media in the three exchangers is calculated as $\pi R^2 \times$ depth × number of units, or 3.14 × (3^2) ft × 7 ft × 3 units = 593.46 ft^3 of media. At 15,000 gr/ft^3, the hardness removal capacity would be 593.46 ft^3 × 15,000 = 8,901,900 gr removal capacity. This would also be stated as 8,901.9 kgr removal capacity. There are three units, so each unit has ⅓ of the total capacity, or 2,967,300 gr.

What is the percent bypass?

Percent bypass is calculated by dividing the finished hardness by the source hardness. In this case, it would be 143 mg/L ÷ 325 mg/L, or 0.44. Expressed as percent, this is 44 percent.

With one high-service pump operating, how many gallons per minute are bypassing the softeners?

One pump operates at 500 gpm; 44 percent of 500 gpm is 220 gpm. This is the flow rate of the water bypassing the softeners. (This also means that the rest of the water, or 280 gpm, must be going through the softeners.)

How long will one unit operate before it needs to be regenerated?

If 280 gpm are going through the softeners, it can be assumed that the flow through one softener, if balanced, is 280 gpm ÷ 3 = 93.3 gpm. Of course, meters would be installed on the influent to the softeners to help the operator balance the flow, which is desirable. A hardness of 325 mg/L is converted to grains per gallon by dividing it by 17.1. This yields a hardness of 19 gpg. Therefore, a unit is being loaded with source water at a rate of 93.3 gpm, each gallon of which carries with it 19 gr of hardness. After calculating the removal capacity of one unit to be 2,967,300 gr, divide that capacity by 19 gpg to determine that about 156,174 gal of water can be softened before exhaustion. At 93.3 gpm, it will take about 28 hours to exhaust the unit (156,174 gal ÷ 93.3 gpm/60 min/hour).

Approximately how much finished water is produced during this time?

This is a tricky, but simple, question. Regardless of the hardness being removed, if all is going well, the plant is operating at 500 gpm (ignoring backwash, plant use, etc.). Therefore, 500 gpm × 60 min/hour × 28 hours = 840,000 gal.

Assume the plant is producing 720,000 gal of finished water per day before plant usage. How many pounds of salt will be used in a 30-day month?

Only 56 percent of that water had to be softened; so in a month, it is calculated (0.56 × 720,000 gal/day × 30 days) that approximately 12.1 mil gal flowed through the exchange units. First calculate the amount of hardness, in grains, that was removed: 12,100,000 gal × 19 gpg = 229,900,000 gr, or 229,900 kgr. Use 0.3 lb of salt per kilograins of hardness removed, so 0.3 lb/kgr × 229,900 kgr = 68,970 lb needed for the month.

How many gallons of saturated brine is this?

At about 2.5 lb/gal, 27,588 gal (68,970 lb × gal/2.5 lb) would need to be used.

Ion-Exchange Residuals

Spent brine used in the regeneration process is typically sent to a sanitary sewer. In coastal regions, water utilities may obtain permits to discharge to the ocean. Other methods that have been used to dispose of brine are discharged to evaporation ponds in arid regions or to brine disposal wells if they exist. Disposal of regenerant brine solutions used in cation radionuclide removal must be approved by regulatory agencies.

Other Softening Processes

There are several additional processes that are sometimes used to soften drinking water: electrodialysis, membrane softening, and split stream treatment (and blending strategies). Electrodialysis is discussed in some detail in chapter 10 because it is used mainly for desalination rather than exclusively for softening. Membrane softening is usually accomplished using specially designed nanofiltration membranes (reverse osmosis membranes also remove hardness). Membrane systems are discussed in detail in chapter 11. The operation and design of membrane softening systems are analogous to other membrane systems.

Split stream and blending are not separate treatment processes but rather are strategies to adjust water hardness where the opportunity exists. Some treatment processes like reverse osmosis and electrodialysis can remove calcium and magnesium almost completely for the water stream. This may result in a water that is "too soft" for the intended purpose. Some systems blend this demineralized water with unsoftened water to provide a water that meets the hardness goals. This is often termed *split treatment*. Another similar approach is to blend hard water with a softer water source before it is distributed. One example where this can be accomplished is in a system that has both surface water (soft) and groundwater (hard). The two are blended and medium hardness water is distributed to customers. Where this is possible, this can be a cost-effective approach.

BIBLIOGRAPHY

APHA, AWWA, and WEF (American Public Health Association, American Water Works Association, and Water Environment Federation). 2012. *Standard Methods for the Examination of Water and Wastewater*, 22nd ed. Washington, D.C.: APHA.

AWWA. 2010. Principles and Practices of Water Supply Operations—*Water Treatment*, 4th ed. Denver, Colo.: American Water Works Association.

Barnhart, R.K. 1986. *Hammond Barnhart Dictionary of Science*. Union, N.J.: Hammond Inc.

Edzwald, J.K., ed. 2011. *Water Quality and Treatment*, 6th ed. New York: American Water Works Association and McGraw-Hill.

Pizzi, N.G. 1995. *Hoover's Water Supply and Treatment*. National Lime Association. Dubuque, Iowa: Kendall-Hunt Publishers.

Rodriguez, M.N. 2000. The Alkalinity Coefficient—A Best Lime Dosage Guide. Unpublished paper contributed by city of St. Louis, Dave Visintainer memo to author.

Chapter 10

Specialized Treatment Processes

Federal and local regulations and certain site-specific conditions may require the use of specialized technologies for water treatment. The removal of regulated contaminants may indicate the use of certain targeted treatment processes. And some of these processes may be necessary in order to produce a product that is aesthetically acceptable to the customer. The descriptions included in this chapter give a brief overview of the most prevalent specialized treatment processes. The references included at the end of the chapter should be consulted for additional detailed information.

AERATION

Aeration, or air stripping, is used to make water stable and used for disinfection, precipitation of inorganic contaminants, and air stripping of volatile organic compounds (VOCs) and nuisance-causing dissolved gases. The most commonly used types of aeration are diffused air, surface aerator, spray, and packed-tower systems (Figure 10-1). These systems remove or minimize the offending contaminants by stripping them out of the water (in the case of gases) or by oxidizing them (e.g., iron and manganese).

Diffused-air or bubble aeration involves contacting gas bubbles with water in order for the gases to transfer to the water. Operators use this technology to strip VOCs from the water, apply ozone, oxidize with O_2, or recarbonate with CO_2. The most common arrangement for this process consists of perforated tubes or porous plates situated at the bottom of the reactor tank. This allows for maximum contact time of the gas bubbles as they rise through the water. The point of application is dependent on the purpose of the process. For example, lime precipitative softening plants that use recarbonation may add CO_2 to the effluent of the clarifiers just prior to filtration, while operators using ozonation for disinfection may add CO_2 to the raw water. The use of ozone technology is covered further in chapter 8.

Surface aerators are used primarily for oxygen absorption and to strip gases and volatile contaminants.

Source: US Filter.

Figure 10-1 Forced draft aerator used to strip organic chemicals from source water

ADSORPTION (GAC)

Adsorption involves the accumulation of a substance at the interface of two phases, such as a solid phase and a liquid phase. An example of this process is the collection of taste- and odor-causing compounds in water on the surface of activated carbon particles. Ion-exchange resins can also be used as interfaces to adsorb *organic* compounds. The use of ion exchange to remove *inorganic* compounds is discussed in chapter 9.

Granular activated carbon (GAC), perhaps the most commonly recognized solid-phase adsorption method used in public water treatment, is a good adsorber because of its surface chemistry and its surface area. In fact, GAC particles are actually honey-combed, with cracks and fissures. This property gives the particle a very large surface-area-to-mass ratio. These fissures are capable of capturing (adsorbing) many small organic molecules over extended periods of time. GAC particles, tailored to adsorb different types and sizes of organic molecules, can be specified for different purposes. GAC with small pore openings can be used to trap low–molecular-weight organics, and the larger pore–sized GAC particles can be used to trap the high–molecular-weight organics. This same property can, of course, allow the GAC to be fouled or prematurely used up if care is not taken.

GAC is made from raw materials such as lignite, wood, peat, and bituminous coal. The manufacturing process involves heating the material to temperatures near 700°C in the absence of air or to 800° to 900°C in steam and CO_2. The process and raw materials used determine the characteristics of the final product. GAC is often used as a cap for granular filters and serves the dual purpose of particulate removal and particulate adsorption.

There are several important concepts that operators must know about GAC. Empty-bed contact time (EBCT) is a design parameter used to predict performance. It is the detention time of the water that flows through a bed at a given flow rate. The higher the flow rate, the lower the EBCT for a fixed bed depth. For source water of a given quality, there is a minimum depth of GAC that will prevent breakthrough of the adsorbate. All GAC beds remove adsorbates in an area called the *mass transfer zone.* The zone must be deep enough to provide sufficient contact time as the water filters down through it. If there is insufficient zone depth or if the design flow rate is exceeded, breakthrough of adsorbate will occur. Therefore, an EBCT that provides adequate opportunity for water/GAC contact is necessary.

Breakthrough will also occur when the bed is exhausted. As the bed removes adsorbate, the available adsorption sites on the GAC become saturated. After a time, the GAC cannot accept any more adsorbate and it must be replaced or reactivated. For source waters with different kinds of adsorbates, it is common for a GAC bed to become exhausted selectively for each. For example, a GAC bed may begin to pass disinfection by-products long before it passes tastes and odors.

Taste-and-Odor Control

The control of disagreeable tastes and odors in public water supplies is normally accomplished by oxidation or adsorption processes. In addition, certain odors caused by volatile gases can be stripped from the water by aeration, as discussed in a previous section.

There are two strategies that can be used to produce water that is free of tastes and odors. The utility can take measures to prevent or reduce the likelihood that offending compounds will reach the source water or, once in the water, the utility can take measures to remove them. The choice may depend on economic factors and/or plant capabilities. Most plant operators would agree that prevention is a better practice than removal after the fact, but it is not always practical.

The test for taste and odor is called the *threshold odor test.* It is an attempt to classify the characteristics, including intensity, of the particular odor that is found in a water supply. The test compares successive dilutions of finished water to odor-free water. The secondary maximum contaminant level (SMCL) for odor is 3 TON (threshold odor number). Operators normally try to keep this value as low as possible, because odors tend to erode consumers' confidence in the water supply. When customers lose confidence, they may seek alternative and less reliable or unsafe supplies of water.

One method for taste- and odor-treatment that is used in conventional treatment plants is to apply a GAC cap to a granular media filter. The detention time is short because the cap is usually only 6–24 in. deep. In a dual-media or multi-media filter, GAC can be added to the top of the filter or it can replace the anthracite layer. Filter flow rates and head loss characteristics must be considered. Several treatment plants have reported a successful implementation of this approach. Some have used this process to remove DBPs or DBP precursors to acceptable levels.

IRON AND MANGANESE REMOVAL

Iron and manganese are naturally occurring minerals found in rocks and soil and are usually found in an insoluble form. However, as rainwater percolates into soils, dissolved oxygen is removed by the decomposition of organic materials. As oxygen levels decrease, the water becomes capable of dissolving iron and manganese. This is why many groundwaters have iron. Manganese is found more rarely, but when it is present, it is usually along with iron.

Iron and manganese do not present public health concerns. They do, however, create considerable aesthetic and operational problems when present in water supplies. As soluble iron and manganese become oxidized, they transform from colorless compounds to turbid yellow and black suspended solids that can stain sinks and fixtures and clog water mains. In some cases, iron or manganese will cause odor problems when they react with other chemicals. Water plants the sources of which contain these minerals are usually equipped with treatment devices that remove the minerals before water is distributed to the public.

Because iron and manganese can cause problems, the US Environmental Protection Agency (USEPA) has set the SMCL for iron at 0.3 mg/L. Customers seem to notice a taste from iron if the level reaches 0.5 mg/L. An SMCL has also been set for manganese at 0.05 mg/L. Manganese stains on laundry are very stubborn, and most operators strive to eliminate manganese altogether.

Iron and manganese in water are controlled at the treatment plant using the following methods:

- Precipitation and filtration
- Ion exchange
- Sequestration

As mentioned in chapter 9, the lime-softening process also removes iron and manganese quite effectively. However, this process is not used solely for iron and manganese treatment but in combination with softening.

Precipitation Processes for Iron and Manganese Removal

In precipitation, soluble forms of iron and manganese are oxidized to convert them to insoluble forms. When completely converted to an insoluble form, they can be effectively settled and filtered. To be oxidized, the iron must come into contact with air or oxygen or the iron and manganese must be chemically oxidized using chlorine, ozone, potassium permanganate, or chlorine dioxide.

Aeration oxidizes iron in water that has a pH greater than 6.5, and aeration is also useful for manganese oxidation when the pH is less than 9.5. Long contact times may be necessary to complete the process. Water can be aerated by forcing compressed air through it or by allowing water to cascade down over trays, effectively allowing for thin sheets of water to be exposed to the air. The trays become coated with iron and manganese hydroxide over time, which promotes further removal of the metals. A reaction tank or vessel is usually provided downstream of the aeration equipment. This tank is designed for a suitable detention period so that complete oxidation of the metals can take place. Detention times of 20 minutes to 1 hour are common.

Chemicals are commonly used for oxidation and can eliminate the need for double pumping and for the reaction vessel. Chlorine, chlorine dioxide, and potassium permanganate are the oxidants of choice. These powerful chemicals can be used to produce insoluble forms of iron and manganese and, in the case of potassium permanganate, can reduce the disinfectant/disinfection by-product levels because no chlorine is added at the front of the plant.

All of these processes require the removal of the iron and manganese floc that is formed. At higher concentrations of the metals, sedimentation is specified, followed by filtration. If pH in the sedimentation basin is low, lime may be used to raise the pH to levels that are more favorable for precipitation. Filtration alone can be used after oxidation if the concentration of the metals is below about 5 mg/L.

Ion Exchange

Ion exchangers can remove iron and manganese, along with other metals, if the proper exchange resin is used. Where hardness is low and there are no precipitates to foul the bed, this process is specified for waters with low iron content. Otherwise, the process is not usually economical.

Manganese Greensand Filtration

This process is commonly used for the removal of iron, manganese, and hydrogen sulfide from groundwater. The manganese greensand (glauconite) is a naturally occurring material that can be classified to produce an excellent filtration media having a sieve analysis of 18 x 60 mesh with a resulting effective size of 0.3–0.35 millimeters (mm) and a uniformity coefficient of 1.60 or less.

To optimize the process, the material is first coated with manganese oxide. This stabilizes the material and imparts a special oxidation-reduction property that is effective for removing iron, manganese, and small amounts of hydrogen sulfide. The filter media must be regenerated when their capacity to remove iron and manganese is reached. Regeneration is accomplished either continuously (CR) or intermittently (IR).

Continuous regeneration is usually used when iron is mostly being removed. An oxidant, either chlorine or potassium permanganate ($KMnO_4$), is fed continuously ahead of the greensand filters. The oxidized iron is captured on the filter sand. IR is used mainly for manganese removal. The greensand is treated with oxidant (usually permanganate) when the filter capacity is reached. This intermittent treatment occurs while the filter is off-line.

Operating conditions may vary and optimization is site specific. However, as a starting point, the pH is usually adjusted so as to be in the 6.2–6.8 range. Alkalinity is ideally >120 mg/L as $CaCO_3$. When iron alone is treated, chlorine is fed at about 1.0–1.2 times the iron concentration. For water containing both manganese and iron, most plants feed chlorine followed by permanganate. Chlorine is fed, in this case, at about 1.0 the iron concentration and 2.0 the manganese concentration. This is followed by a small permanganate feed that is controlled by the faint pink color in the water just before the greensand filter.

Backwashing the greensand filters must be performed periodically to remove captured iron and manganese particles and to restore the filter rate capacity. To enhance filter performance and improve backwash efficiency, the greensand filters are often capped with anthracite filter media. It is common for the filters to have an 18-in. anthracite cap over 18–30 in. of greensand (depending on the regeneration method and the amount of iron and manganese being removed).

Operating a greensand process can be tricky and often careful monitoring of pH, chlorine, permanganate, iron, manganese, and hydrogen sulfide levels is necessary. Sufficient training and instruction regarding proper operation of the chemical feed systems and filter operation are necessary to achieve optimum results.

Sequestration

If the source water is low in iron and manganese (<1.0 mg/L combined) and contains no dissolved oxygen, sequestration may be used to treat for iron and manganese. Sequestration is the process by which iron and manganese are kept in solution by adding chemicals, such as polyphosphates or silicates. In this way, the iron and manganese are not removed but rather prevented from being oxidized and made insoluble. Bacterial slimes sometimes occur in systems that sequester because the iron is still available. Care must be taken to maintain adequate chlorine residual to minimize this problem.

FLUORIDATION OF WATER

Fluoride is a mineral that occurs naturally in all water sources, even in the oceans. It is never encountered in the free state but rather in combined form.

When fluoride levels in water are held constant at about 1.0 mg/L, the incidence of dental cavities is reduced. Because of this fact, public water supplies are usually fluoridated to optimum levels. In 2006, the Centers for

Disease Control and Prevention (CDC), Atlanta, Georgia, stated that 184 million Americans, or 69 percent of those served by public water supplies, consume fluoridated water.

Dental caries, the most prevalent chronic disease of human beings, is most widespread in children. Tooth decay reaches a peak in adolescence and diminishes in adulthood. It is for this reason that fluoridated water has the greatest effect on the teeth during the formative years. Fluoride is absorbed on the teeth and bones and actually is incorporated into the crystal of the tooth enamel. When teeth are forming, this helps to strengthen the enamel by forming a layer that is more impervious to the bacteria that cause decay.

Fluoride levels above the optimum can cause mottling of teeth from dental fluorosis. This condition, which creates brittleness in the teeth and moderate to severe discoloration, was first noticed in 1901 in Naples, Italy, as reported by the US Public Health Service. Water purveyors must control the levels of fluoride ion concentration in their supply. The USEPA has set a maximum contaminant level (MCL) for fluoride at 4 mg/L to prevent crippling skeletal fluorosis. The Department of Health and Human Services and USEPA have recommended that the optimum level of fluoride in drinking water is 0.7 mg/L.

In some communities, there may be controversy surrounding the practice of fluoridation. Most water plant operators are in compliance as directed by local health officials. Assistance with response to community concerns can be obtained from the CDC.

Chemicals Used in Fluoridation

Fluoride is the negative ionic form of fluorine, a gaseous element. Fluoride compounds can be found in all soils, water supplies, plants, and animals in varying amounts. Fluorine is a reactive element and therefore is always found in combination form, never in the free state. Fluoride-containing minerals are used by industry to produce fluoride compounds, the most prevalent being fluorspar, cryolite, and apatite. From these minerals, the three most common fluoride-adjustment compounds are generated: sodium fluoride (NaF), hydrofluosilicic acid (H_2SiF_6), and sodium silicofluoride (Na_2SiF_6). These three compounds are by-products of the phosphoric acid fertilizer industry, and their availability is dictated by agricultural production. Sales of fertilizer have a direct effect on the manufacture and availability of fluorides for water treatment.

Sodium Fluoride

Sodium fluoride (NaF) is a white odorless powder or crystal. It has a specific gravity of 2.79, a molecular weight of 42 (sodium = 23, fluorine = 19), and a solubility of 4 g/100 mL, or 4 percent in water. When added to water, NaF dissociates into sodium and the needed fluoride ions. NaF can be purchased in quantities that are very pure, usually near 97 to 98 percent.

A dosage of 19 lb NaF into 1 mil gal of fluoride-free water will produce a 1-mg/L F^- solution. The percent of available fluoride in NaF is 45 percent (19 ÷ 42).

The calculation for estimating the dosage of 1 mg/L F^- when using NaF (97 percent purity) is

$$19 \text{ lb/mil gal NaF} \times ([1 \text{ mg/L}]/8.34 \text{ lb/mil gal}) \times 0.97 \text{ purity} = 1.0 \text{ mg/L } F^-$$

Sodium Silicofluoride

Sodium silicofluoride (Na_2SiF_6) is a white odorless crystalline material. It has a specific gravity of 2.679, a molecular weight of 188 (2 sodiums = 46, 1 silicon = 28, 6 fluorines = 114), and a solubility ranging from 0.44 to about 2.5 percent. Na_2SiF_6 can be purchased in purities of 98 percent or higher.

A dosage of 14 lb Na_2SiF_6 into 1 mil gal of fluoride-free water will produce a 1 mg/L F^- solution. The percent available fluoride in Na_2SiF_6 is about 61 percent (114 ÷ 188).

The calculation for estimating the dosage of 1 mg/L F^- when using Na_2SiF_6 (98 percent purity) is

$$14 \text{ lb/mil gal } Na_2SiF_6 \times (1 \text{ mg/L} \div 8.34 \text{ lb/mil gal}) \times 0.61 \text{ available } F^- \times 0.98 \text{ purity} = 1.0 \text{ mg/L } F^-$$

Hydrofluosilicic Acid

Hydrofluosilicic acid (H_2SiF_6) is a straw- or amber-colored, transparent, fuming, corrosive liquid. It has a molecular weight of about 144 (2 hydrogens = 2, 1 silicon = 28, 6 fluorines = 114) and can be purchased as a liquid in drums or bulk at 23 to 35 percent.

A dosage of 46 lb of 23 percent (H_2SiF_6) into 1 mil gal of fluoride-free water will produce a 1-mg/L F^- solution. The percent of available fluoride in (H_2SiF_6) is about 79 percent (114 ÷ 144).

The calculation for estimating the dosage of 1 mg/L F^- when using (H_2SiF_6) (23 percent purity) is

$$46 \text{ lb } (H_2SiF_6) \times 1 \text{ mg/L} \times 0.79 \text{ available } F^- \times .23 \text{ purity}/8.34 \text{ lb/mil gal} = 1.0 \text{ mg/L } F^-$$

Note that water has no "fluoride demand" and therefore a contact period, such as for chlorine, is not needed. Some fluoride ion will, however, be lost in sedimentation and filtration (usually 10 to 30 percent), so the optimum feed point for fluoride is at postfiltration. The fluoride injection point should be as far away as possible from any chemical feeds that contain calcium, such as lime. It is prudent to check the fluoride concentration on plant tap water daily and to check the feed system hourly while in operation.

Any overfeed of fluoride is a public health concern. All medical and emergency agencies should be notified, especially kidney dialysis centers, if over-fluoridation occurs. In addition, notify the CDC.

Feed Equipment

There are three accepted methods for feeding supplemental fluoride to a water supply:

- Fluoride saturator (Figure 10-2)
- Dry chemical feeder (Figure 10-3)
- Chemical solution feeder (Figure 10-4), using either H_2SiF_6 or a solution made from the dry chemicals

Feed equipment is selected based on many factors, including plant size, availability and type of fluoride source, cost, and technical expertise of personnel. Larger facilities seem to prefer liquid storage and feed systems. The operator is directed to AWWA Manual M4—*Water Fluoridation Principles and Practices*, for more information.

Fluoride chemicals are added to water as liquids but may be measured as either liquid or solid. When solid chemicals are dissolved into solution before feeding, the strength of the solution must be controlled, usually by feeding a predetermined amount of chemical for a specific duration. Two types of dry-feed machines exist: volumetric and gravimetric. The volumetric feeder delivers a predetermined volume of dry chemical within a given time; the gravimetric feeder delivers a predetermined weight of chemical within a given time and is more accurate.

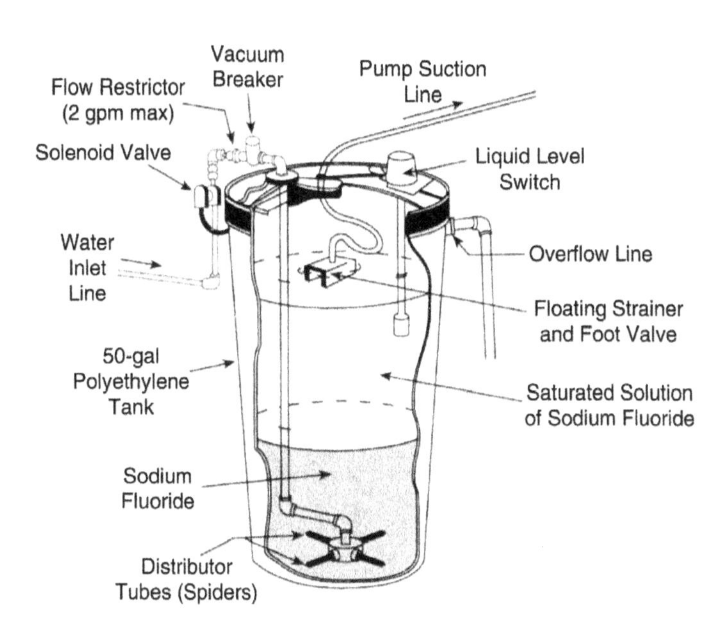

Figure 10-2 Solution fluoride saturator

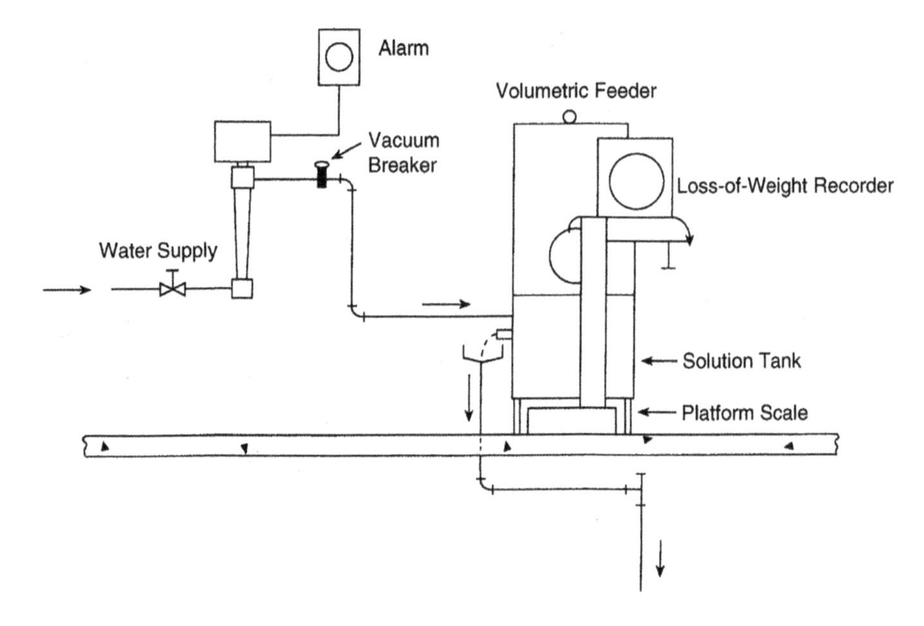

Figure 10-3 Dry feed installation with volumetric feeder for fluoride

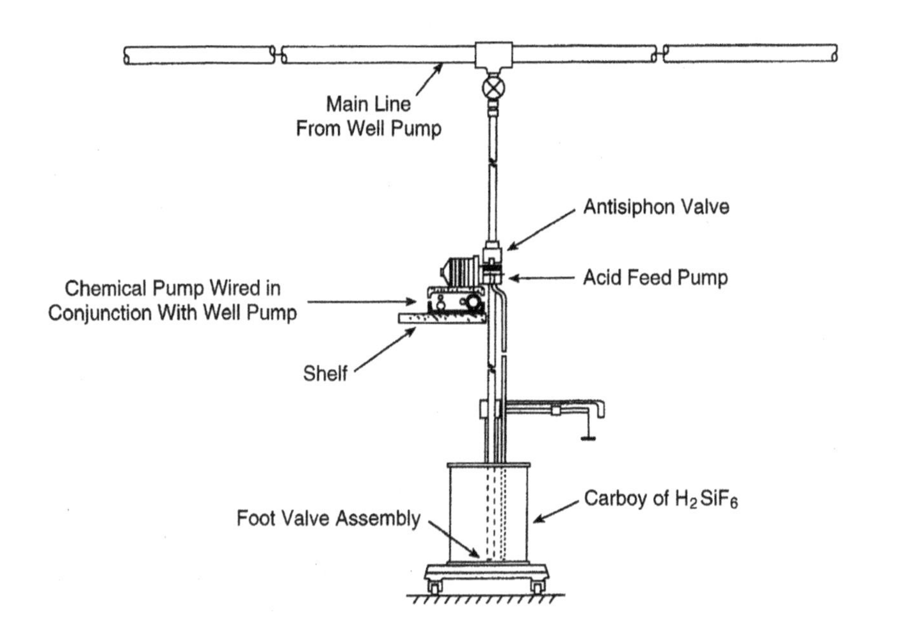

Figure 10-4 Acid feed installation for fluoride

Safety Precautions

Fluoride dust from NaF and Na_2SiF_6 can expose operators to higher-than-optimal levels of fluoride and can irritate the nasal and mucous membranes. Fluoride acids such as H_2SiF_6 can burn the skin. The following precautions are advisable when handling these chemicals:

- Avoid breathing fluoride dust. Dust masks should be issued to all who work around fluoride dust.
- Wear rubber gloves and aprons when handling dust.
- Wash away any dust from the skin immediately.
- When handling hydrofluosilicic acid, wear acid-proof gloves, apron, and face shield. Wash these items after use.
- Do not allow persons with open cuts or sores to handle any fluoride acids or dusts.
- Post safety rules in the vicinity of the chemical storage and handling area.
- Label fluoride containers as poisonous and never reuse the containers for other purposes.
- If accidentally ingested, give the victim a 1 percent solution of calcium chloride, a glass of saturated lime ($Ca(OH)_2$) water, or milk to drink. Flush the chemical out of the eyes with warm water. Get medical attention immediately.

Example Dosage Calculation

Problem: A water plant treats 7.21 mgd and wishes to maintain a fluoride ion level of 1.1 mg/L in the plant effluent. If the raw water source contains 0.36 mg/L natural fluoride ion, how many gallons of 24.3 percent H_2SiF_6 will be needed in the month of September? The acid used is 10.5 lb/gal.

Answer: H_2SiF_6 has 79 percent available fluoride ion.

$$1.1 \text{ mg/L desired} - 0.36 \text{ mg/L available} = 0.74 \text{ mg/L } F^- \text{ needed dosage}$$

$$0.74 \text{ mg/L } F^- \times (8.34 \text{ lb/mil gal } F^-/1 \text{ mg/L}) \times 7.21 \text{ mgd} \times (30 \text{ days/month}) = 1{,}334.92 \text{ lb } F^-/\text{month}$$

$$1{,}334.92 \text{ lb } F^-/\text{month} \div (.243 \times .79) = 6{,}953.8 \text{ lb } H_2SiF_6/\text{month}$$

$$(6953.8 \text{ lb } H_2SiF_6/\text{month}) \times (1 \text{ gal}/10.5 \text{ lb}) = 662.26 \text{ gal/month}$$

ACTIVATED ALUMINA FLUORIDE REMOVAL PROCESS

Some source waters contain fluoride is excess of the regulatory limits and, thus, must be reduced. Both surface and groundwaters may be subject to fluoride removal necessity. Fluoride can be removed by several processes including: RO, electrodialysis, adsorptive media, and anion exchange. Activated alumina (AA) is a selective ion exchange material that is often the method of choice for fluoride removal.

Activated alumina (AA) is a granular filter media that is primarily aluminum oxide (Al_2O_3). It has been activated by heat and washing with sodium hydroxide (caustic soda). Fluoride ions in water are attracted to the surfaces of the AA. This is a pH-sensitive process where the optimum is a pH of 5.5. Fluoride can be removed to below 0.5 mg/L at this pH.

The capacity of the AA filters varies depending on the concentration of fluoride in the feed water. As the capacity of the filter is reached, the media can be regenerated by washing with a caustic solution. The wash waste can be recycled and the AA media can be used again to remove fluoride from the water.

Arsenic (III and V) is also removed from water by AA. If present in the water, arsenic will compete with fluoride for adsorption on AA. This is not often a problem because arsenic is usually present in very low concentrations when compared to fluoride. Therefore, it does not materially affect the capacity of the AA for fluoride removal. A modification of the AA treatment process can be employed in the rare case where arsenic is present in high concentration. Arsenic held on the AA requires higher concentrations of caustic for regeneration than does fluoride.

A typical treatment system consists of two down-flow pressure vessels in series. Fluoride removal is accomplished in the first vessel initially. As more water is treated, the fluoride-containing area of the AA in the vessels migrates down to the second vessel. The first vessel is then removed from service and regenerated. The second vessel continues to process water containing fluoride. When the first vessel regeneration is completed, it is returned to service in the lag position. In this way, continuous treatment is accomplished.

ADSORPTIVE MEDIA

Fluoride, arsenic, uranium, and other anions are often removed using adsorptive media. Specific materials used can be activated alumina, modified activated alumina, iron sorbents, and other proprietary adsorptive materials. The process operates similarly to the activated alumina process described above. The pH or other operating details may be different from the optimized AA process for fluoride removal. The system manufacturer should be contacted for operating specifics. Extensive pilot testing should be conducted to verify the optimum operating conditions.

ELECTRODIALYSIS DEMINERALIZATION

Dissolved charged ions migrate through semipermeable membranes due to an attraction to electrically charged electrodes (Figure 10-5). Electrodialysis (ED) is able to remove most dissolved ions from water and is an approved process to treat nitrate, nitrite, barium, selenium, and total dissolved solids.

Typical treatment systems require pretreatment with scale inhibitor, chlorine residual of about 0.5 mg/L, and a cartridge filter (10–20 μm). Sometimes air stripping precedes the dialysis modules to remove any hydrogen sulfide.

The ED process relies on the fact that most dissolved ions in water are either positively or negatively charged. Selective semipermeable membranes allow either anions or cations to pass through while blocking those with the opposite charge. This results in a separation of the ions from the body of the water flowing through the system and concentrating the ions in the waste stream. An illustration is shown in Figure 10-5.

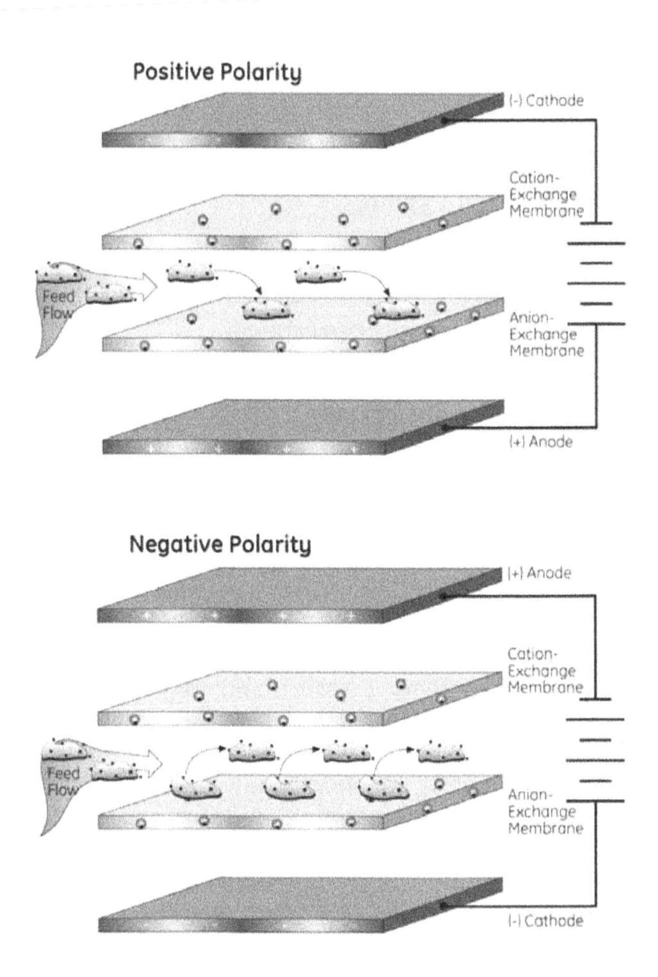

Figure 10-5 Schematic of electrodialysis

Electrodialysis reversal (EDR) is similar to ED but the polarity of the electrodes is reversed regularly. This releases the ions from the electrodes and into the waste stream. This practice reduces scaling and increases the life of the membranes. Summaries of operating conditions and advantages of EDR/ED follow:

1. Typical operating conditions of ED/EDR
   - Turbidity < 2 ntu
   - Iron < 0.3 mg/L (Fe^{+2})
   - Manganese < 0.1 mg/L (Mn^{+2})
   - Hydrogen sulfide < 1 mg/L
   - SDI < 15
   - Free chlorine continuous 0.5 mg/L tolerate up to 20 mg/L for short time
   - TOC < 15 mg/L
   - pH 2–11
   - TDS up to 8,000 mg/L
2. Some advantages of ED/EDR
   - Minimal fouling or scaling
   - Low pressure
   - Long membrane life
   - Tolerate nonionic species like silica
   - Low chemical usage for pretreatment

PRECOAT FILTRATION (DIATOMACEOUS EARTH)

Diatomaceous earth (DE) is the skeletal remains of microscopic diatoms or algae. This material is mostly silica and inert so it is suitable to use to filter water. The process is approved for drinking water treatment.

The process uses DE to strain particulates from water and normally does not use coagulant chemicals. A thin layer of DE coats filter leaves or septum that is a plastic or wire mesh covering a hollow collection channel. The DE coating (precoat) is formed by a recirculating slurry of DE. There are two main types of DE filter systems: pressure and vacuum. Both have advantages and disadvantages.

Untreated water is applied to the precoat filter (usually about ⅛ in. thick) with a small amount of DE added (called *body feed*). Particulates in the water are removed on the filter surface, and this causes pressure to build (head loss). When the terminal head loss is reached, the filter is taken out of service for cleaning. High pressure water spray on the filter cake is used to remove the accumulated material. This waste material is discarded and the process is repeated starting with the filter precoat with fresh DE.

DE filtration is used for raw water with a maximum turbidity of 10 ntu. The filter loading rate can be between 0.5 and 2.0 gpm/ft^2. DE filtration is very effective for removing *Giardia* and *Cryptosporidium*. Dissolved substances and color are not effectively removed by DE filtration.

CORROSION AND SCALING CONTROL

The control or minimization of corrosion in water pipes and distribution systems is an important operational task. Corrosion is the deterioration of a pipe or fixture due to its contact with water. The more corrosive the water is, the more likely that deterioration will take place. When pipes and fixtures corrode, they put materials into the water that may be harmful to the public (lead, cadmium), may stain fixtures (iron, zinc, copper), may reduce the carrying capacity of mains and destroy hot water heaters, or may produce sediments that harbor bacterial slimes, which in turn can cause tastes and odors and additional corrosion.

Corrosion occurs when all components of an electrochemical cell are present. These components include an *anode*, a *cathode*, a *connection* between the anode and cathode that allows electron transport, and an *electrolyte solution* that will conduct ions between the anode and cathode. If any one of these components is missing, corrosion will not take place. During the corrosion process, metal ions dissolve at the site of the anode. The electrons released in this process migrate to the cathode. Positive ions, in turn, migrate to the anode. This migration is the system's attempt to maintain an electrically neutral solution. The rate of migration affects the amount of metal that dissolves and is influenced by several factors. Knowledge and control of these factors, which are described in the following section, can help the operator keep corrosion to a minimum.

Corrosion Factors

The common factors that can affect corrosion are described in the following paragraphs.

Dissolved oxygen. In general, as the concentration of dissolved oxygen increases, the rate of corrosion increases.

Total dissolved solids. Water with a high total dissolved solids (TDS) concentration conducts the flow of electrons. Pure water is not a good conductor. As the TDS of water increases, the rate of corrosion increases.

Alkalinity and pH. Alkalinity and pH affect the rate of chemical reactions in water. A decrease in the corrosion rate is seen as the pH and alkalinity increase.

Temperature. An increase in water temperature can increase the rate of corrosion.

Flow velocity. Flow rates in pipes can positively or negatively affect corrosion rates. Where water is corrosive, a higher flow rate can bring dissolved oxygen to the surface of the pipe more readily and therefore increase the corrosion rate. However, when the water has been stabilized with corrosion-control chemicals (lime, phosphates), a high velocity can decrease the corrosion rate by allowing for faster deposition of protective films.

Type of metal. Metals are not usually found in pure form in nature. Rather, they exist as ores or other impure compounds. Corrosion of metals can be thought of as nature's way of bringing the metal back to a natural state. To add to the problem, certain metals naturally corrode more easily than others. If metals of high activity are combined with metals of low activity, the rate of corrosion between the two is increased. For example, when a copper service connection is coupled directly to an iron main, corrosion will take place; copper is much less active than iron, and iron corrodes more easily. Table 10-1 presents the galvanic series; in this series, each metal is anodic to, and may be corroded by, any metal below it. Table 10-1 illustrates that placing mild steel next to wrought iron would produce a slower rate of corrosion than would placing mild steel next to copper.

The technique of cathodic protection uses this principle. When two dissimilar metals are brought together in a water solution, the most active metal becomes the anode and will corrode. Water operators often put magnesium alloy caps on iron pipe bolts. The caps become the sacrificial anode and are corroded. The bolts are thus protected at the expense of the caps.

Lead and Copper Rule

Much of the lead and copper found in drinking water is leached from household plumbing that corrodes the water. The problem of toxic metals leaching from pipes, which occurs in drinking water, has been addressed by the USEPA in the Lead and Copper Rule (LCR). The health effects of lead and copper in water are well documented, and water suppliers are required to minimize the rates of corrosion of these metals. The LCR provides for periodic sampling from customer taps for the analysis of these two metals. Water that has been allowed to sit still in the pipes for at least 6 hours is sampled, and the values for lead and copper are listed in order of magnitude. The rule requires that 90 percent of lead levels be less than 15 µg/L, and 90 percent of the copper levels be less than 1.3 mg/L. If suppliers cannot meet these levels, treatment techniques that provide for corrosion control are required. Refer to chapter 1 for further discussion of the LCR.

ION EXCHANGE

Groundwater is known to commonly contain high amounts of minerals and other natural substances that are present in the subterranean rocks and soil. Frequently high levels of hardness (calcium and magnesium) and dissolved radioactive ions are encountered and must be removed. Although precipitative softening (lime-softening) can be used to effectively remove hardness, ion exchange is often used instead for smaller flows from many groundwater systems. Ion exchange softening is discussed in chapter 9.

Radioactive ions, usually encountered in groundwater, are generally highly charged and, therefore, are ideal candidates for ion exchange removal processes. Regulated radionuclides susceptible to ion exchange removal include: Radium 226 and 228, beta and photon emitters, are usually present as cations (positively charged). The ion exchange process used to achieve their removal is similar to the ion exchange softening process.

Table 10-1 The galvanic series for metals used in water systems

Corroded End (Anode)	
Magnesium	**Most Active**
Magnesium alloys	
Zinc	+
Aluminum	Corrosion Potential
Cadmium	
Mild steel	
Wrought iron	
Cast iron	
Lead–tin solders	
Lead	
Tin	
Brass	
Copper	
Stainless steel	-
Protected end (cathode)	**Least Active**

Anion Ion Exchange

Some radionuclides are present in water as anions. Special test methods must be used to identify the radionuclide and determine the predominant species that is present. The most common radionuclide anion is uranium. Uranium is often present as a $UO_2(CO_3)_3{}^{4-}$ion (depending of pH the $UO_2(CO_3)_2{}^{2-}$ ion may also be present) in water. Therefore, an anion exchange process can be very effective for its removal. The main differences between cation and anion exchange processes are that an anion exchange resin is used and although sodium chloride is often the regenerant, it is the chloride (Cl^-) that is being exchanged. The disposal of concentrated regenerant for uranium is also a concern as it is with other radionuclides. Refer to chapter 9 for a discussion of ion exchange to understand the concepts of bed volume (BV), exhaustion, and regeneration.

A strong anion exchange resin is used for uranium removal. In many cases, the resin beds are able to effectively remove uranium for more than 200,000 BV. After the bed is exhausted, the bed resin can be disposed (care must be exercised not to exceed the radioactive limits for disposal). If the bed is regenerated instead, the run length is limited (often to 10,000–50,000 BV) so that the used regenerant meets disposal limits.

The concentrations of other anions (like nitrate, nitrite, or organic tannins) may also be removed by anion exchange. Depending on the conditions and amounts, these may compete with uranium and reduce the effective BV throughput. It may, however, be advantageous to plan on removing other anions in conjunction with uranium ion exchange treatment.

BIBLIOGRAPHY

AWWA. 2010. Principles and Practices of Water Supply Operations—*Water Treatment*, 4th ed. Denver, Colo.: American Water Works Association.

AWWA. 2004. Manual M4—*Water Fluoridation Principles and Practices*. Denver, Colo.: American Water Works Association.

AWWA. 2011. M58—*Internal Corrosion in Water Distribution Systems*. Denver, Colo.: American Water Works Association.

AWWA. 2013. M56—*Nitrification Prevention and Control in Drinking Water*. Denver, Colo.: American Water Works Association.

AWWA. 2005. M53—*Microfiltration and Ultrafiltration Membranes for Drinking Water*. Denver, Colo.: American Water Works Association.

AWWA. 1988. M30—*Precoat Filtration*. Denver, Colo.: American Water Works Association.

Chowdhury, Z.K., R.Scott Summers, G.P. Westerhoff, B.J. Leto, K.O. Nowack, C.J. Corwin, L.B. Passantino, Technical Ed. 2012. *Activated Carbon: Soultions for Improving Water Quality*. Denver, Colo.: American Water Works Association.

Edzwald, J.K., ed. 2011. *Water Quality and Treatment*, 6th ed. James K., ed. New York: American Water Works Association and McGraw-Hill.

Pizzi, N.G. 1995. *Hoover's Water Supply and Treatment*. National Lime Association. Dubuque, Iowa: Kendall-Hunt Publishers.

USEPA. *Drinking Water Treatment Wastes*, Radiation Protection. http://www.epa.gov/rpdweb00/tenorm/drinking-water.html

Chapter 11

Membrane Systems

Membrane systems, which can be used to treat water, are a rapidly developing technology and one that is finding increased use in the water supply community (Figure 11-1). Most contaminants found in water supplies can be removed using membrane technology. As the unit cost for installing membranes decreases and drinking water regulations become more stringent, it is likely that more water utilities will explore the use of membranes.

The most common membrane processes in use today are reverse osmosis (RO), nanofiltration (NF), ultrafiltration (UF), and microfiltration (MF). RO is used primarily to treat brackish water or seawater, while the other processes are used to soften water or to remove disinfection by-product precursors, turbidity, and pathogens.

Membranes work by providing a barrier against the passage of suspended, colloidal, or dissolved species found in source water. In general, the cost of membrane treatment increases as the size of the species being removed decreases.

REVERSE OSMOSIS

RO is based on the principle of *osmosis*, which is a phenomenon that occurs when water can flow through a membrane, but certain ions and molecules cannot. The process diffuses molecules through spaces in the membrane's molecular structure. An example is that of two saltwater solutions of unequal concentration and separated by a membrane that gradually become equal in concentration. Water flows from the weaker of the two solutions to the more concentrated one, thereby diluting it. The water flows across the membrane because of an exerted pressure called *osmotic pressure*. RO works by applying a greater pressure on the other side of the membrane to force the flow of water in the opposite direction. Figure 11-2 illustrates this process.

Because RO is able to reject ions such as calcium, magnesium, chloride, and sulfate, very pure water can be treated using this process. RO is used to remove contaminants and to soften and/or desalinate water. The process membranes are capable of removing most single-valent (ions such as sodium) dissolved ions. The operating pressure depends on the chemical

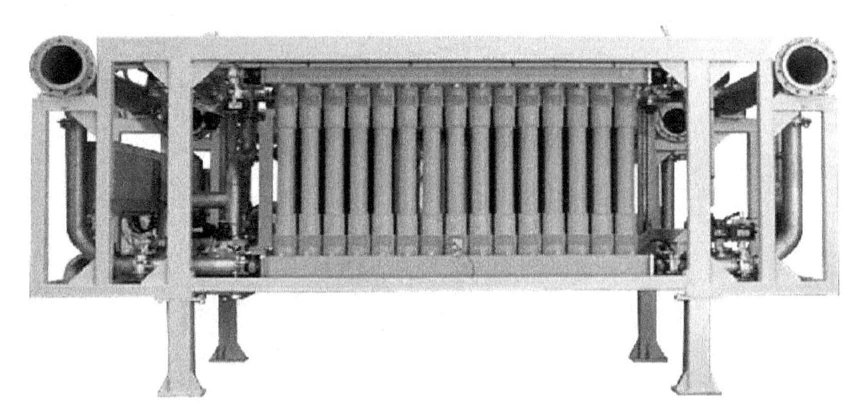

Source: US Filter.

Figure 11-1 Continuous microfiltration unit

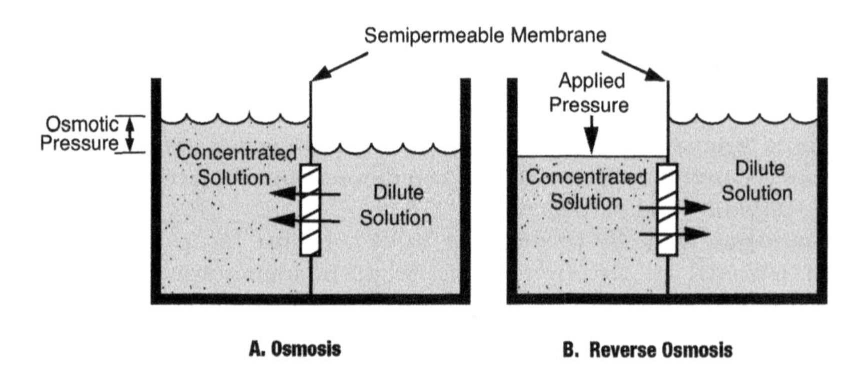

Figure 11-2 Osmosis and reverse osmosis processes

characteristics of the feedwater. Generally, the higher the TDS, the more pressure is required to force the water through the RO membranes. Pressures range from a few hundred psi for brackish water to more than 1,000 psi for seawater.

The membranes used in the RO process are manufactured from materials like cellulose acetate and polyamide-composite (thin film composite or TFC). Each material has certain operating characteristics, such as the ability to reject salt, a pH range at which it works best, resistance to degradation from chlorine or other oxidants, and susceptibility to biological attack.

When new, membranes perform at their peak. As operating time accumulates, the membranes will compact and foul, which affects performance. Compaction gradually closes the pores of the membrane, and the rate of compaction increases with higher operating pressures and temperatures. Fouling occurs when scales from the incoming feedwater deposit on the membrane. Calcium and silica, along with iron and other metal oxides, will foul membranes. For this reason, water high in scaling content is usually pretreated before passing through RO membranes. In some cases, fouled membranes can be restored with an acid wash; otherwise, they are replaced when irreversible fouling occurs. Feedwater pretreatment may require turbidity reduction, iron and manganese removal, pH stabilization, microbial control, chlorine removal, and hardness reduction.

Fouled membranes present an operational concern as water productivity decreases and higher pressures are needed to produce an effluent. Most operators rely on the following rules of thumb for an indication that fouling has occurred:

- The passage of salt through the membrane increases by 15 percent
- The pressure drop through the unit increases by 20 percent
- Feed pressure requirements increase by 20 percent
- Product water flows drop or increase by 5 percent
- Fouling or scaling is evident

The RO operation will produce an amount of reject water, which can range from 20 to 50 percent of the feedwater to the unit. The amount of reject water is dependent on the number of stages in which the membranes are configured and the feed pressure. Reject water poses two potential problems for the operator: there must be sufficient quantity of water source to supply up to two times the amount of water needed by the system and the reject water must be disposed of properly. Where water is scarce or costly, RO systems may be problematic. Where other minerals overtax sewer systems, the reject water from RO may not be acceptable. Both of these potential problems place an emphasis on efficient operations.

Although most RO systems are operated using programmable logic controllers, operators must monitor the system for operational problems and efficiency. Operators should check raw water quality, including the chemical and biological makeup of the source. The pretreatment acid system must be inspected frequently, and the pH must be maintained at a predetermined level. Cartridge filter drains should be exercised periodically, and the head loss across the filters should be recorded. Any post-treatment systems should also be monitored for proper operation.

The major problems experienced by operators of RO systems are premature fouling and flux decline. Flux, or compaction, can usually be predicted by the manufacturer of the unit. If operators notice a more rapid rate of flux decline, they should consult with the manufacturer. Because chlorine can be especially hard on some membranes, operators need to be wary of that problem. Certain membranes are resistant to chlorine, and it may be recommended to maintain a residual in the feedwater in some systems.

NANOFILTRATION

NF uses membranes with a pore size of about 0.001 mμ (1 μ = 0.001 mm). The NF molecular weight cutoff is in the 1,000–100,000 Dalton range and operating pressure is up to 150 psi. NF is gaining in popularity as a potable water treatment device because it can remove hardness and reduce salinity of impaired water supplies. Nanofiltration can be used to remove disinfection by-product precursors such as humic acid.

ULTRAFILTRATION

UF generally uses membranes with a pore size of about 0.002–0.1 mμ. The molecular weight cutoff is in the 10,000–100,000 Dalton range and the operating pressure is 30–100 psi. It therefore can remove colloids and other large molecular substances such as organic compounds. It does not remove hardness or salt but has the advantage of using less energy than the other systems described previously.

The principle advantage of UF is its ability to remove *Giardia* and *Cryptosporidium*, as well as viruses that may be present in surface water supplies.

MICROFILTRATION

The MF process generally operates with an effective pore size of about 0.03–10 mμ. The systems have molecular weight cut off of greater than 100,000 Daltons and feedwater operating pressure of about 15–60 psi. It can be used as a pretreatment process for other membrane units or to remove very fine particles in industrial applications, such as the electronics industry (Figure 11-3). Unlike the high-pressure RO and NF units, this lower-pressure system (including UF) must be backwashed frequently, which poses operational constraints on operators.

This technology is finding use in the direct filtration process because it can remove small particles that inhibit disinfection. In fact, an uncompromised system (one in which all membrane elements are intact) can achieve absolute removal of *Giardia* and *Cryptosporidium* organisms. Unlike UF, MF does not remove all viruses present in the surface water supply.

Some existing conventional water treatment plants, such as the Kenosha, Wisconsin, facility, have opted to increase plant capacity by placing MF units in parallel with the conventional train. Kenosha's 40-mgd facility was increased to 60 mgd (approximately).

Membrane Integrity Testing

Low-pressure membranes (MF and UF) are commonly used for drinking water treatment. The membrane surface, although quite uniform, may have imperfections or may develop gaps during operation that could allow the passage of some contaminants through the system. Concern has been amplified where the membrane is the primary barrier in the treatment process.

Integrity testing of membranes is a necessity to provide assurance that the system is operating as intended to provide a positive barrier to the passage of pathogens. This testing can be direct, indirect, and can be conducted

Source: Siemens Industry Inc.

Figure 11-3 Submerged microfiltration unit

while the system is in operation, online, or during maintenance operation, off-line. Some systems use several testing methods to provide added confidence to the integrity of the system.

Direct test methods include: pressure decay test, diffusive air flow, bubble point test, acoustic sensor test, liquid–liquid porosimeter test, the air hold test, and the binary gas integrity test. Indirect tests include: particle counting, particle monitoring, turbidity monitoring, and surrogate challenge tests. Several of these test methods have been adapted to online monitoring.

Most of the integrity tests now available are for the low-pressure membrane systems. Similar methods for high pressure (NF and RO) are under development. Indirect methods are being used routinely. Research is being conducted to establish integrity tests that are most useful for these systems.

New or refined test methods are under development. Many of the integrity tests are specific to the membrane type or even the manufacturer. Each system has its own requirements and tests are employed that are the most suitable to the situation. This is an area that will continue to evolve and improve. The goal is to have a continuous online monitoring system to ensure membrane integrity.

POST-TREATMENT

Treatment of membrane product water consists of different unit operations for RO and NF systems. Primarily, the treatment schemes may include aeration, pH adjustment, disinfection, and filtration. There may also be a need for alkalinity recovery and elimination of hydrogen sulfide, if present. Note that there is little or no demand for chlorine after membrane treatment, and so operators will need to dose at the level they need for the distribution system. Alkalinity recovery, needed because of the destruction of alkalinity in the acid step, is beneficial to the system in terms of corrosion control and LCR compliance. Product water stability should be controlled to reduce corrosion.

FUTURE CONSIDERATIONS

Conventional plants are considering converting to MF or UF facilities in phases as these alternatives are becoming more cost competitive. Membrane integrity can be monitored so closely that failure in a single module can be determined; this is quite important to the reliability and acceptability of the systems.

Extensive membrane system research will continue to provide lower cost alternatives. Improvements in membrane concentrate disposal and reuse will make adoption of these technologies even more attractive to many water suppliers.

BIBLIOGRAPHY

AWWA. 2010. Principles and Practices of Water Supply Operations—*Water Treatment*, 4th ed. Denver, Colo.: American Water Works Association.

AWWA. 2005. M53—*Microfiltration and Ultrafiltration Membranes for Drinking Water*. Denver, Colo.: American Water Works Association.

AWWA. 2011. M61—*Desalination of Seawater*. Denver, Colo.: American Water Works Association.

Edzwald, J.K., ed. 2011. *Water Quality and Treatment*, 6th ed. New York: American Water Works Association and McGraw-Hill.

Liu, C. 2012. *Integrity Testing for Low Pressure Membranes*. Denver, Colo.: American Water Works Association.

Wolfe, T.A. 2001. MW Harza Global. Internal communication to author. October.

Chapter 12

Testing and Laboratory Procedures

REASONS FOR TESTING

Water plant operators take samples every day. Water in the plant and the distribution system is sampled frequently for two reasons: for compliance testing to ensure the water is safe for human consumption and to measure the efficiency of the treatment process. Some samples, such as those for disinfectant residuals and turbidity, are tested on-site and continuously; others, such as those for disinfection by-products or metals, may be sent periodically to a laboratory for analysis. The size and resources of the individual utility will affect the choices that the operator makes regarding sampling.

SAMPLING

Reliable test results depend on proper sampling procedures and techniques. Samples are supposed to accurately represent the quality of the water being sampled. Consequently, consideration must be given to the way the sample is collected, the volume taken, any storage requirements for the sample, and sampling locations and frequencies.

Sample Collection

Improper sampling is a common cause of error in water quality analysis. Because the results of an analysis can only show what is actually in the sample, the sample must have the same content as the water from which it was taken. Water laboratory analysts refer to this as a *representative sample*, and it is defined as being one of two types: grab sample or composite sample.

Grab samples are single-volume samples collected at one time from one place. An operator taking a 100-mL sample from a distribution system for bacteriological analysis is taking a grab sample. Similarly, an operator taking a few milliliters of settled water for turbidity analysis is taking a grab sample. As such, these samples represent the quality of water in those locations only at the time that the sample was obtained. For this reason, grab samples are best used when the quality of the water is not expected to change significantly or where changes take place slowly. A grab sample might not give an accurate representation of the quality (over time) of a flowing river that is subject to intermittent pollution from upstream wastewater plants or chance rainfall events. The frequency of grab sampling is determined by studying the history of changes in water quality, so that adequate representation of the quality can be made.

A composite sample consists of multiple grab samples taken from the same location at different times and mixed together to make one sample. Time-composite samples are made up of equal-volume grab samples taken at different times. An example of a time-composite sample is 100-mL samples of backwash water taken at 1-minute intervals for the length of the wash and mixed together to measure total solids. Flow-proportional composite samples consist of different volumes of sample taken from the same location at different times; the volume collected each time is dependent on the flow rate of the water at the sample location. When composite samples of raw water in a water treatment plant are needed and the raw water flow rate varies over the day, a proportional composite sample is sometimes taken because it more accurately represents the water quality over time.

Composite samples allow the laboratory analyst to determine the average concentration of a constituent over time, without having to perform the analysis every time a sample is taken. Automatic samplers can be used to take samples at night, which enables the chemist who comes to work in the morning to perform the analysis. There are drawbacks to composite samples. The high and low points, or ranges, of a contaminant cannot be known because they are averaged. Also, some constituents, such as pH, dissolved gases, chlorine residual, and temperature, begin to change immediately after the sample is taken and therefore should not be composited. Bacteriological samples, such as those taken for coliform analysis, must always be taken as grab samples because the number of bacteria in the samples begins to change immediately after sampling.

Sample Volume

The volume of sample will vary according to the testing procedures used and often is set by regulations. For example, coliform compliance sampling requires a 100-mL grab sample. The laboratory analyst will provide instructions for sample volume.

Sample Storage

Glass and carbonate plastic sample containers are most often used in water utilities, but glass is avoided where there is a greater chance of breakage, such as in shipping. Also, glass is not used when sampling for fluoride because fluoride will adhere to the glass and not provide accurate results. Plastic can also be problematic because certain organic chemicals can permeate the plastic, and therefore the organic chemicals will not be detected in the subsequent analysis. Some types of plastics may actually release organic chemicals into the sample and so invalidate the results.

Samples for total trihalomethane and haloacetic acid analyses are taken in glass vials, usually colored brown to filter out sunlight, and shipped in coolers. Often, sample containers have preservatives, quenching agents, or pH-adjustment chemicals in them to facilitate sample preservation and/or storage, which extend the shelf life of the sample, allowing more time for analysis. This in turn allows the analyst to collect all the necessary samples and then perform a single analysis, greatly reducing errors that may otherwise occur during calibration. It also is more efficient and less costly to perform one calibration for multiple samples than it is to calibrate for each individual sample.

Regulated parameters have specific directions for sampling, preservation, and storage before testing. These requirements are listed in the regulatory agency compliance regulations. Some agencies have adopted the USEPA requirements and some have separate, additional requirements. Many, including USEPA, have referred to *Standard Methods for the Examination of Water and Wastewater.* Consult the regulatory agency for requirements in the utility's location.

Sample Location and Frequency

Selecting representative sampling points and determining sampling frequency are important steps in attaining meaningful water quality data. Samples are usually collected from three areas within the treatment system: the raw or source water, the plant, and the distribution system. Because sources and treatment schemes differ from system to system, no guidelines are given in this handbook. Local regulatory agencies usually provide a minimum frequency for sampling, and most operators take samples with a greater frequency than is required.

In-plant sample locations should be chosen to allow for the measurement of the performance of each unit process. Operators should take care not to sample immediately after chemical addition but rather at a point where proper mixing has been achieved. Sampling of combined filter effluent turbidity should take place before post-treatment chemicals are added to avoid measuring added turbidity, such as from lime or phosphates.

Distribution system samples are intended to measure the quality of the water that is delivered to customers at various points in the system. In general, samples should not be taken from customers' taps. The exception is for lead and copper testing and for water quality complaint sampling. The inlet and outlet of storage tanks are good sampling sites because they give information about the effects of storage. The direction of flow should be noted when sampling, i.e., is the tank filling or emptying at the time of sampling.

Bacteriological sampling and the associated chlorine residual testing are regulated. Frequency of sampling is based on population served.

QUALITY ASSURANCE/QUALITY CONTROL

Standard Methods for the Examination of Water and Wastewater defines quality assurance (QA) as "a definitive plan for laboratory operation that specifies the measures used to produce data of known precision and bias." It defines quality control (QC) as a "set of measures within a sample analysis methodology to assure that the process is in control."

A QA program of a laboratory consists of a QA manual, written procedures, work instructions, and records. For example, an organizational chart that lists the training, capabilities, and responsibilities of each lab analyst would be part of a QA program. Quality control is also part of the overall QA program.

In a QC program, each analyst demonstrates his or her capability to obtain acceptable results for an analysis. This is demonstrated by analyzing a set of "blind" standards. Analysts are given samples that contain a specific amount of a chemical. The analyst runs the samples, and the results are reported to an agency that knows the correct levels of chemicals in the samples. The agency then rates the laboratory and the analyst on their performance. This type of testing is commonly called a *performance evaluation* or PE. PEs are typically run twice a year and determine if a laboratory is "certified" to run the analysis.

Another type of continuous laboratory evaluation is performed through the use of "known" standards. Known standards are samples that contain a predetermined amount of a chemical. These standards along with each set of samples are run by an analyst. This allows an analyst to determine how well a particular set of analyses has been performed in each analytical trial. Power fluctuations or other daily hazards may affect laboratory equipment, and the evaluation of known samples will let an analyst determine if a particular run is satisfactory. Sometimes these known samples are referred to as *standards*. For example, known turbidity standards are run before each set of filter turbidity grab samples is run, thus allowing an operator to determine if the instrument is properly measuring turbidity.

A final set of samples is also used to determine data quality. These samples are referred to as *spiked samples*, *laboratory control samples*, or *laboratory fortified matrix (LFM) samples*. Spiked samples are made in the laboratory by adding a known quantity of the chemical being measured to the collected samples. The analyst splits one sample into two aliquots and then adds a precise amount of the chemical being analyzed to one of the aliquots. The two aliquots are run and the results compared. The sample to which the chemical was spiked should have a level of chemical equal to the known addition plus the results of the unaltered sample. This type of sample is important because it allows the analyst to determine if some other chemistry in the sample is interfering with how the instrument measures the contaminant. Regulations and reference to *Standard Methods for the Examination of Water and Wastewater* determine the frequency of this type of QA sampling. Different analyses require different numbers of QA samples; this will affect the number of samples an operator may be required to take for any given parameter.

Another way to evaluate laboratory performance is by reviewing historical analytical performance. This is known as *quality control*. The results of known samples and spiked samples are graphed and reviewed for trends over time. This is important because the efficiency of analytical equipment can decrease over time, or an analyst can improve his or her performance with experience. For example, a decrease in the amount of a known sample result over time can indicate that a probe or detector needs to be replaced.

The importance of laboratory QA/QC to overall plant operations cannot be overstressed. It is critical that test results be reliable because they are a measure of how a plant is performing and a measure of the safety of the water the plant is producing. Performance must be measured accurately, and the accuracy of analysis depends on sample quality and analytical quality. All laboratories should strictly adhere to the laboratory practices described in this section and be willing to produce QA/QC information as requested.

Common Water Utility Tests

Following is a brief description of some of the more common parameters that water plant operators frequently measure, including their significance. Where applicable, cross-references to other chapters are given.

Alkalinity

The buffering capacity of the water is measured by its alkalinity. This test can determine the concentration of carbonate (CO_3^{-2}), bicarbonate (HCO_3^-), and hydroxide (OH^-) alkalinity. These measurements are useful to determine the corrosive nature of the water in combination with other factors and to optimize the lime–soda ash softening process.

The test involves careful titration of a measured quantity of water with a standard 0.02 *N* sulfuric acid solution to pH end points of 8.3 and 4.5 (these are indicated by color change indicators or by using a pH meter). The carbonate, bicarbonate, and hydroxide alkalinity are then calculated from the results.

Carbon Dioxide

Water that contains high concentrations of free carbon dioxide (CO_2) can cause the consumption of lime when using this method of softening. Also, carbon dioxide is a factor affecting corrosion.

The test usually used is to titrate a measured water sample with a standard solution of sodium hydroxide. The end point is signified by a change in color for phenolphthalein or a pH of 8.3.

Chlorine (free or total)

For water plants that use chlorine for disinfection or oxidation, this is one of the most important tests performed by operators. After any addition of chlorine, a measurement should be taken routinely. This test will verify the correct dosage and reveal changes that may affect plant performance and the safety of the water supply. Many plants use both free and total (or combined) forms of chlorine in their processes.

The test most often used is the N,N diethyl-1,4 phenylenediamine sulfate (DPD) color test. The DPD (either for free or total chlorine testing) is added to a water sample, and the intensity of the color indicates the amount of chlorine present in the sample. Most plants use a digital read-out colorimeter to give an accurate result. Some color comparison portable test devices are used as well; however, they are not always accurate. Another test method is amperometric titration. This test is usually performed in the laboratory. A special meter is used to determine the end points when titrating a measured sample with a standard PAO solution. This method is very accurate and is capable of determining many chlorine species that may be present. There are several other chlorine residual test methods (*Standard Methods for the Examination of Water and Wastewater*) that can be considered.

Both the DPD color test and amperometric titration are used in online instruments for continuous chlorine monitoring. As with any instrument, these devices must be calibrated and checked frequently to ensure accuracy.

Chlorine demand can be determined using the residual test method. This measurement can be used to predict residual chlorine over a specified time. Jar testing apparatus may be used for this test. A sample is taken and the chlorine residual is measured immediately. After a specified time, the residual is measured again and the difference is the demand. Care is needed to duplicate the conditions (light, temperature, holding time) that are of interest.

Chlorine Dioxide

See chapter 8 for a discussion of chlorine dioxide disinfection, generation, and operation. Also, chapters 9 and 10 include some information regarding ion exchange for iron and manganese removal and greensand filter operation. Chlorine dioxide residual is limited to 0.8 mg/L (USEPA MRDL) and a by-product, chlorite, is also regulated with an MCL of 1.0 mg/L.

The test methods for chlorine dioxide are similar to those for chlorine. There are DPD and amperometric test methods, but some of the test conditions have been modified to yield chlorine dioxide–specific results. Also, there is an ion chromatography method that requires trained technicians in a certified laboratory.

Coliform

Coliforms (see chapter 8) are a group of bacteria that produces gas bubbles in lactose or lauryl tryptose broth at 35.5°C within 24 to 48 hours. They are considered indicator organisms—their presence may indicate the presence of other more harmful bacteria and organisms. Because total coliforms are easier to analyze in the average water utility laboratory than the actual disease-causing micro-organisms, total coliform testing is used in place of more tedious, more expensive testing for these other organisms.

Most laboratories use one of two methods to test for coliforms. One method is the membrane filter technique, in which water samples are passed through a 0.45-μm filter. The filter paper is fine enough to trap bacterial particles as the water passes through. The filter paper is then placed onto a growth medium (such as M-endo) and incubated. Any bacterial colonies that are present will grow in size, be visible to the naked eye, and can be counted after a period of time (usually 24 hours).

Another method is the MMO-MUG (minimal medium) technique. This method allows for inoculation of water sample bottles with powder. The substance in the bottle will feed any total coliform that may be present and produce a color change during incubation.

Samples for coliform testing are always collected in sterile bottles and in quantities sufficient for testing (Figure 12-1). Most bacteriological samples require a minimum of 100 mL (approximately 4 oz) for analysis. Samples should be analyzed the same day as they are collected but can be refrigerated for eight hours prior to analysis. Coliform samples are taken in the plant at various stages to test for process efficiency; they are also taken in the distribution system for regulatory compliance. The number of samples that must be taken in the distribution system is a function of the population served. Many water treatment plants also take coliform samples of the raw or source water.

Source: Conneaut, Ohio, Water Department.

Figure 12-1 Autoclave used for sterilization. Bacteriological equipment must be sterilized in steam under pressure.

Conductivity

This test (more correctly named *specific conductance*) measures the ability of water to conduct electricity. This is an indirect measure of the ions or minerals in the water. It is sometimes used to estimate the total dissolved solids content due to its ease of measurement and the availability of inexpensive mobile field instruments.

A conductivity meter is connected to an electrode, and this is immersed in a water sample. An instrument setting to match the sample temperature is often used to adjust for extremes. The conductivity reading is in microsiemens/centimeter (µs/cm).

Cryptosporidium

Cryptosporidium is a parasite regulated under the Long-Term 2 Enhanced Surface Water Treatment Rule (LT2ESWTR). It is a regulated pathogen, and its measurement is necessary to ensure adequate treatment depending on the occurrence in the untreated water supply.

The method involves filtration, specialized separation, and identification methods. This test can only be performed by laboratories approved for this method. Plant operations personnel will use the test results to determine compliance with the LT2ESWTR.

Disinfectant By-Products (DBPs)

Several organic compounds are created when chlorine is used in water treatment. Two groups of these by-products are regulated: trihalomethanes (THMs) and haloacetic acids (HAAs). Compounds in both groups have been classified as probable carcinogens.

The test methods for both groups involve procedures that must be performed by trained analysts in approved laboratories. Compliance testing is, therefore, not usually performed by plant operating personnel. Online instruments are available for continuous monitoring but the instruments should be checked periodically by the same approved laboratory procedures. Operator-performed screening tests can be used to indicate levels for plant monitoring, but these methods are not approved for compliance testing.

Dissolved Oxygen

The amount of dissolved oxygen (DO) in the source water may be an indicator of the condition of the lake or reservoir being used. Lack of oxygen may be a predictor of water quality problems due to anoxic conditions. The DO concentration in the treated water may be a factor contributing to the rate of corrosion.

The DO test involves the use of a dedicated meter and specific sensor probe. Also, there are color methods where the intensity of the color from the addition of the indicator is a measure of the concentration present in the water.

Fluoride

Fluoride reduces tooth decay. Fluoride may be naturally occurring or added during water treatment as directed by local health authorities. The optimum amount of fluoride needed to provide health benefits depends on several factors. There is a maximum MCL of 4.0 mg/L for fluoride in drinking water.

There are both instrumental and colorimetric tests for fluoride in water. Operations personnel usually use the color methods. The color indicator is added to the water and the color intensity measures the amount of flouride present.

Giardia lamblia

Giardia lamblia is a parasite found in untreated water supplies, and it is pathogenic. It is regulated by a treatment technique where disinfectant concentration and contact time (CT) are prescribed to ensure adequate inactivation. Separate requirements are specified for chlorine, chloramine, ozone, and UV.

The method involves filtration, specialized separation, and identification methods. This test can only be performed by laboratories approved for this method. Plant operations personnel will use the test results to determine compliance with the SWTR.

Hardness

Hardness is mostly attributed to the amount of calcium and magnesium in the water. Softening treatment plants must carefully monitor hardness as a process control. Most other water system regularly monitor for this parameter as a basic measure of water quality.

There are several methods for hardness testing. Although hardness can be determined by separately analyzing the calcium and magnesium levels, most water systems use a color titration method for total hardness results. A measured sample is titrated with a standard solution (EDTA) to a color change end point.

Inorganics (heavy metals)

Some metals are toxic in high amounts (such as cadmium or chromium) and others are not (such as iron or manganese). There are MCL regulations for many heavy metals and secondary limits for those that can cause aesthetic concerns. Most regulated metals are included in the primary drinking water regulations and some have special rules such as lead and copper.

The test methods for metals are tremendously varied. Instrumental and color methods predominate. Only approved methods can be used for compliance testing. Operations personnel may elect to use unapproved screening methods for process monitoring. The more precise methods require trained analysts and must be performed by approved laboratories.

Iron and Manganese

These metals are often grouped together because they are similar in their role in water quality and often occur together. Both cause staining of fixtures and are not generally toxic at normally encountered concentrations. Treatment processes have been developed to remove one or both of these substances. Secondary standards (nonenforceable) are established for both.

There are several test methods for both of these metals. Many treatment plants use a color method for process control. These methods are typical where an indicator is used, and the intensity of the color is measured to determine the amount present.

Lead and Copper

These metals are grouped together because of the USEPA regulatory rule requirements. Both substances are toxic at elevated levels and, thus, they are regulated. Generally, these contaminants are somewhat unique in that they are not often found in source water but are the result of dissolution from piping components (see the Lead and Copper Rule description in chapter 1). Therefore, testing is mostly performed in the distribution system rather than the treatment plant.

The test methods vary from instrumental laboratory procedures to simple field-test color methods. The laboratory methods require trained analysts and are not usually conducted by operations personnel. Screening tests using colorimetric methods are often employed by plant operators for process control.

Microbiological Organisms

Many microbiological organisms are found in untreated water. Bacteria, viruses, protozoan, algae, and plankton are examples of the multitude of possibilities. Some of these are pathogenic; others can produce toxic substances. The array of consequences is very broad. Some are regulated in drinking water. The coliform group of bacteria is used to indicate the possibility of contamination. Another useful bacteria indicator is the heterotrophic plate count (HPC). This is a general bacteria population measurement that may indicate the presence of other pathogens.

Test methods are as varied as the organisms. Many involve growing colonies on selective nutrients and counting the number from a measured sample. Others may be examined by microscope and identified by experts. Some chemical by-products are analyzed by instrumental methods capable of detecting minute amounts. Generally, training analysts must perform most of these procedures. The exceptions are the coliform tests that have been developed using color changes to both detect and enumerate these bacteria.

Nitrate, Nitrite, and Ammonia

These three inorganic nitrogen compounds are often encountered in water supplies. Contamination from agricultural activities is often the source in surface water, but another significant source is wastewater discharges. Nitrate and nitrite are regulated contaminants with enforceable MCLs. Ammonia is not regulated in drinking water but is a compound that may encourage the growth of nitrifying bacteria. Ammonia is often associated with the chloramination process.

Several test methods are available for these substances. Operational control testing is usually conducted using colorimetric methods. Special precautions are needed when testing for free ammonia in the presence of chloramines in order to ensure accurate results.

Ortho- and Poly-Phosphate

Phosphate can be naturally occurring in surface water where it may be the result of urban runoff or waste discharges. Ortho-phosphate and poly-phosphate are often used as corrosion inhibitors in water supplies. These substances are not considered toxic, but the amount is carefully monitored to ensure optimum effectiveness.

There are instrumental methods for phosphate that require approved laboratories and trained analysts. Most operational control testing uses colorimetric tests. Selecting the correct test may require knowledge of the type of phosphate used for corrosion control.

pH

This is one of the most common tests performed in water treatment and drinking water monitoring. The pH value (chapters 5, 9, and 10) is an indicator of the acidity or alkalinity of the water. There is not an MCL for pH in drinking water. However, there is a secondary standard range of 6.5–8.5.

It uses a scale from 0 to 14, with the midpoint of 7 being neutral, i.e., the acidity and alkalinity are balanced. Below 7, the acidity of the water predominates; above 7, the alkalinity of the water predominates. With each unit increase or decrease, the concentration or intensity changes tenfold.

For example, for the pH to change from 5 to 6, the acidity must decrease by a factor of 10. The pH of water is significant because it affects the efficiency of chlorination, coagulation, softening, and corrosion control. Also, pH testing can provide early warning of unit process failure. For example, the addition of alum to the rapid-mix stage should produce a predictable drop in pH. If it does not, a malfunction of coagulant feed could be indicated.

Samples for pH should be collected in glass or plastic containers and analyzed as quickly as possible. Samples should not be agitated because dissolved carbon dioxide could be liberated, which will change the pH.

A pH meter is used for the test in combination with a suitable probe. The meter must be periodically calibrated using a known standard. The pH value for a water sample may change while standing due to a change in temperature or exposure to air. Therefore, measurements are usually taken immediately upon sampling.

Radiologic Substances

Contamination from radiological substances is more common in groundwater than surface water. Several of these substances have MCL regulations. There are numerous possible radioactive isotopes that may be encountered; therefore, screening water supplies for general radioactivity (gross alpha and gross beta activity) may be prudent before employing specific substance testing.

Testing for radioactive substances often requires specialized equipment and procedures. Approved laboratories are required for compliance testing and analysts with specific training are needed. Test results for radioactivity are sometimes expressed as pCi/L (picoCuries/L) but, depending on the substance, μg/L or mrem/yr may be used.

Solids—Settleable, Dissolved, Suspended

The solids contained in a water supply are all of the substances that are not H_2O. These can be inorganic, organic, volatile, nonvolatile, suspended, settleable, or dissolved. There are not MCLs for solids content, but there is a secondary standard for total dissolved solids (TDS) of 500 mg/L.

The test method involves taking a measured sample and heating it to remove the water and then weighing the remaining residue (solids). The temperature used for heating can define the volatility of the residue. Also, filtration before heating can be used to give a dissolved result. TDS is the most used solids measurement in drinking water. TDS is a gross measure of the inorganic content of the water (because organic substances usually are a minor part of the total).

Synthetic Organic Chemicals

This is a large group of organic compounds that is regulated and has MCL standards. Many of these are pesticides and herbicides that for the most part are not currently manufactured in the United States. Most of these compounds are probable carcinogens and, therefore, have MCLG limits of 0. These must be tested according to the requirements of the Primary Drinking Water Regulations.

The test methods for this group of compounds involve complex chemistry instrumental procedures. Only approved laboratories with trained analysts can perform these tests. A few of these substances have intricate rapid test methods that are not approved for regulatory compliance. These methods may be suitable for screening surveys or other occurrence evaluations.

Taste and Odor

Consumers often react to taste and odor as their only way to evaluate the quality of the drinking water. A disagreeable response may result in a poor perception so it is imperative that the water utility seek to provide water that is pleasing to its customers. There is no regulatory MCL for taste or odor, but there is a secondary standard MCL for threshold odor number (TON) of 3.

There are several tests for taste and odor. One is the flavor profile analysis (FPA). This test involves a trained panel to routinely evaluate the water. Several utilities use this method with good results. However, most utilities find this method to be labor intensive and instead use the older TON method. This method enlists a panel to smell water sample of various dilutions. The consensus of the dilution where an odor was detected is the TON for the sample. Although this test can be a bit subjective, it can also be useful to detect odor problems and to assess the effectiveness of treatment.

Temperature

This is probably the test most frequently performed on drinking water. There is no MCL for temperature. Differences in temperature can indicate probable water quality problems. Also, water supply changes are often linked to temperature.

A thermometer is used and the water must be tested at the sample location. The result is usually expressed in °C because this scale is used for many other test procedures.

Total Organic Carbon (TOC)

This parameter is a nonspecific measure of the organic content of the water. There is no MCL for TOC, but the Disinfectants and Disinfection By-products Rule (DBPR) uses TOC measurements to determine if precursor removal is needed to comply with the rule. Enhanced coagulation or other means may be required to reduce the TOC.

The test uses a TOC instrument either in the laboratory or online. Calibration is required by analyzing known standards and adjusting the instrument to provide an accurate result. Specific training is necessary to calibrate and use the instrument to provide the best results.

Volatile Organic Chemicals

Several volatile organic chemicals (VOCs) are regulated and have MCLs. Although these compounds are volatile, it is possible that water under pressure may contain these substances, and when the pressure is released, the compounds could be inhaled.

Test methods for this class of compounds involve careful sample collection (zero headspace). The samples are purged with an inert gas, and the vapor is analyzed by gas chromatograph (instrument). This test must be performed in an approved laboratory by trained analysts. Operational personnel may assess the results and compare them to regulatory standards.

Jar Testing

Jar testing is perhaps the least understood test but the most useful process tools available to operators. A few simple ideas and techniques need to be mastered. When the plant staff becomes efficient in this process control strategy, they can rapidly respond to treatment upsets. This is especially important with the more stringent turbidity requirements of the IESWTR.

Jar testing can be used for many tasks, such as screening new coagulants and polymers, coupon testing for corrosion control, biological spiking experiments, etc. The jar testing described in this handbook focuses on coagulant dosage control and oxidant demand. In cases where turbidity breakthrough in a water treatment plant has occurred, an examination of the records often indicates that the staff did not respond adequately to a sudden change in source water turbidity or a sudden source water demand for oxidant. In these cases, the staff simply did not know what dosage of chemicals to use to counteract the changes because they could not perform jar testing.

Jar testing requires a working knowledge of stock solutions. The following text explains how to prepare stocks for alum, alum–polymer blends, ferric chloride, chlorine, and potassium permanganate. It also shows how to perform jar tests with these stocks to determine coagulant dosage and chlorine or permanganate demand. The jar test apparatus, shown in Figure 12-2, is the device most commonly used. It consists of a series of six 2-L square jars with stirrers. A common shaft operates the stirring mechanism ensuring that each jar is mixed identically. Operators should follow the procedures outlined in the manufacturer's recommendations for operation.

Alum

Alum dose is usually expressed in milligrams per liter as dry basis. If liquid alum is used, the product is shipped to the plant as approximately 50 percent alum (8.3 percent Al_2O_3). Commercial alum is approximately 642 mg/mL dry basis. To make a working stock solution for jar testing, pipette 15.6 mL into a 1-L volumetric flask and dilute to the 1-L mark with distilled water. 642 mg/mL × 15.6 mL is approximately 10,000 mg, so this stock is 10,000 mg/L, or 10 mg/mL dry-basis alum. Every milliliter of this stock placed into a 2-L beaker for jar testing would produce a dosage of 5 mg/L alum dry basis.

Alum–Polymer Blend

Normally, the dosage for this product is expressed as gallons per million gallons or parts per million. This is based on the simple idea that one gal of coagulant added to each million gallons of raw water is 1 ppm. This expression negates the requirement to know the exact proportions of alum and polymer in the coagulant.

Note that parts per million and milligrams per liter are not the same. Parts per million is a volume-per-volume relationship, while milligrams per liter is a weight-per-volume relationship. As a rule of thumb, a dosage of 18 ppm would deliver approximately 10.6 mg/L dry-basis alum. The calculation is: if 18 ppm of alum–polymer blend is fed into the system, that is the same as 18 gal of product per 1,000,000 gal of water. Alum–polymer blend

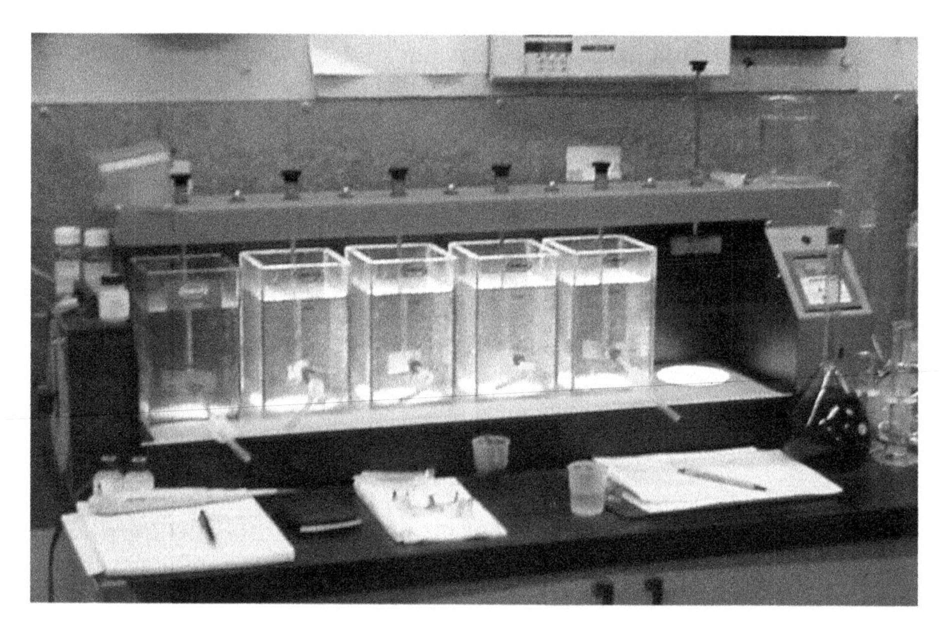

Source: Cleveland, Ohio, Division of Water.

Figure 12-2 Jar testing equipment is used to determine proper coagulant dosage and for process trial testing. Square jars provide better mixing characteristics than round jars.

weighs about 10.9 lb/gal (it may vary), so 18 gal × 10.9 lb/gal = 196.2 lb/mil gal. 196.2 lb/mil gal/8.34 = 23.5 mg/L of liquid alum–polymer blend. Because roughly 45 percent of this is dry alum, 45 percent × 23.5 = 10.6 mg/L dry-basis alum. Because more dry-basis alum is usually required for TOC removal than is needed for turbidity removal, operators should resist lowering blended dosages.

To make a working stock, place 10 mL of the commercial product into a 1-L volumetric flask and fill to the mark with distilled water. This is a 10-parts-per-thousand stock. Each milliliter of this stock placed into a 2-L jar is a dosage of 5 ppm.

Ferric Chloride

Ferric chloride dosage is usually expressed as a milligrams-per-liter product. The material generally comes to the plant as a liquid with a stated percentage that can vary according to manufacturer. Most NSF-certified ferric chloride products are limited to a dosage of 250 mg/L; however, most plants use far less.

To make a working stock solution of ferric chloride from the dry powder, dissolve 2.93 g into 1,000 mL of water and mix thoroughly. The resulting stock is 1,000 mg/L or 1 mg/mL. One milliliter of this stock added to a 2-L jar is a dosage of 0.5 mg/L dry ferric chloride. To make stock solutions from the liquid product, it is best to consult the supplier.

When it is desirable to compare alum and ferric salt coagulants for enhanced coagulation comparisons, make a working stock solution from the liquid product (assuming 38 percent ferric chloride, $FeCl_3$, specific gravity 1.4), transfer 10.27 mL to a 1-L flask and fill to the mark with laboratory-grade water. This solution is 10 mg/mL ferric chloride. One milliliter of this solution put into a 2-L Gator jar will be a dosage of 5 mg/L. The calculation is: 38 percent is 38 g/100 g, and when multiplied by the specific gravity, it yields a concentration of 0.532 g/mL. At 0.532 g/mL × 10.27 mL, a solution is produced that is 5.5 mg/L ferric chloride, which is equivalent to a dosage of 10 mg/L alum. Because the enhanced coagulation requirements are written in 10-mg/L increments of alum, this will allow the operator to make easy comparisons.

Chlorine

Gaseous chlorine (stored as compressed liquid) is used for disinfection and for chemical oxidation. The gas is fed through chlorinators equipped with site tubes graduated in pounds per day or in pounds per million gallons. Dosage for chlorine is calculated using the formula:

$$\text{pounds fed/million gallons}/8.34 = \text{milligrams per liter}$$

The impurities in the water will exert a demand for chlorine, so that the residual amount in the water will be less than the amount fed. Other factors, such as temperature and detention time, will also affect the demand for chlorine. For this reason, operators should know how to test for chlorine demand.

To make a working chlorine stock using commercial bleach, make an approximate 2,000-mg/L chlorine stock by adding 40 mL bleach to a 1-L volumetric flask and fill to the mark with deionized (DI) water. Verify the strength of this stock by adding 1 mL of it to 1 L of DI water. This should test as 2 mg/L free chlorine residual because DI water has no chlorine demand to it. Whatever it tests at, record and use that number when spiking jars for chlorine.

To perform a chlorine demand test, 2 L of raw water should be placed into a square jar and stirred at a slow speed. While stirring, add 2 mL of the stock to the square jar and time the process. After 15 minutes, obtain a sample from the jar for chlorine analysis. The dosage minus the residual is the demand. For example, if the stock solution prepared as above tested as 1.8 mg/L, then 4 mL of that stock placed into 2 L of raw water would produce a dosage of 3.6 mg/L. If, after 15 minutes, the residual chlorine of the jar is 2.0 mg/L, then the demand is 3.6 to 2.0, or 1.6 mg/L.

Note that when performing a simple chlorine demand test, no other oxidants should be added to the jar. Also, commercial bleach has a pH greater than 10 and may increase the pH of the water in the jar.

Potassium Permanganate

Potassium permanganate is used as an oxidant for iron and manganese control, some organic precursor control, and for taste-and-odor control. Permanganate will enhance the coagulation process as it begins to oxidize some of the constituents in the source water. It is thought that the continuous use of this chemical will further reduce the need for chlorine at the front end of the plant. Dosage of permanganate is calculated like any other dry chemical: pounds fed divided by million gallons, in turn divided by 8.34 will provide a dosage as milligrams per liter. Potassium permanganate has the unique quality of a color change mechanism, which signals to the operator its effectiveness and end point. A pink color is associated with water freshly dosed with permanganate. This color changes to straw or yellow at the end point. Operators can use this knowledge as an aid for dosage and demand predictions. Permanganate reactions are dependent largely on pH.

Like chlorine, there is a demand for permanganate. Therefore, operators can use jar-testing techniques to calculate this demand.

Make a 1-mg/L stock solution each week by weighing 1 g potassium permanganate into a 1-L volumetric flask and diluting it to 1 L. This is a 1-mg/mL working stock. Obtain a raw water sample taken before the chemical addition point for potassium and pour it up to the 2-L mark in a Gator jar at the jar test station. Add 1 mL of stock permanganate and agitate for 5 minutes at slow speed on the stirrer. This is a 0.5-mg/L dose. Add the appropriate amount (current operational amount) of stock alum solution and continue to agitate for another 20 minutes. Observe color change during that time.

Sample the settled water from the beaker and run a free chlorine residual test on it. Multiply the result by 0.89—this is the milligrams per liter of permanganate left in the sample. Subtract this from 0.5 mg/L (dosage). The result is the amount that should be fed. For water of very little demand, adjustments may be necessary.

Coagulant Jar Testing

Preparation for Jar Testing

Note: Following are generic procedures for jar testing. Because each water treatment plant is different, the procedures will have to be customized (calibrated) by the operators to fit the particular system. See chapter 5 for a discussion of jar test calibration.

The purpose of jar testing is to simulate full-scale plant operations. Successful jar testing depends on the ability of the operator to simulate the conditions of the water plant as closely as possible. Important considerations are those of proper dosage, proper chemical addition sequence, proper mixing times, and maintenance of operating temperatures. Also important is the need for good sampling and analysis techniques and application of the data produced to the full-scale operation.

Operators should determine the times and sequences of the jar testing procedure using knowledge of their particular plant. These sequences will provide for test results of higher quality if they are accurate. AWWA Manual M37—*Operational Control of Coagulation and Filtration Processes* is an excellent resource for helping the operator determine times and sequences that should be used for testing. The main reason that jar testing fails to produce suitable results is that operators make the mistake of assuming that detention times for settling that are found at plant scale must be duplicated at the bench scale. In reality, bench-scale settling times should be adjusted, or scaled down, and M37 shows how to do that. Most importantly, the sedimentation basin overflow rate must be known because it determines the length of time that floc should be allowed to settle in the jars before sampling. Because overflow rates are a function of plant flow, jar test results can take on new meaning each time an operator makes a change in pump rate.

Starting with fresh stocks, fill weigh boats or pipettes with the proper amount of chemicals and place them next to each jar before proceeding. If using a programmable jar test apparatus, it should be preset for the times that are being used. Manual machines will have to be controlled by an operator at each step change.

Fill each Gator jar with 2 L of fresh raw water just prior to testing. The raw water should be obtained from a point in the process before any chemical addition. Alternatively, if permanganate is being used and the testing being performed is for turbidity removal results only, raw water with the permanganate already in it can be used.

Jar Test Procedure

A generic procedure for jar testing for coagulant-dosage control is provided in this section. Plant operators should alter these procedures and times to suit their particular needs. Consult AWWA Manual M37 for a more detailed jar testing procedure.

1. Begin the testing by setting the mixers to moderate speed. Operators should have calculated detention times of each sequence of their plant.
2. Add the permanganate (if desired) and allow mixing for a time equal to the travel time (baffling factor adjusted) to the rapid mix. If applicable, add chlorine stock.
3. Add coagulant and set mixers to full speed for the duration of the detention time that has been chosen. Be sure to add the coagulant to all jars simultaneously.
4. Turn down the speed of the mixers to simulate flocculation energy for the duration chosen.
5. After flocculation, turn the mixers to the lowest speed possible to simulate flow through the settling basin.
6. After 5 minutes, samples can be taken for analysis. The most common analysis is the test for turbidity (Figure 12-3).

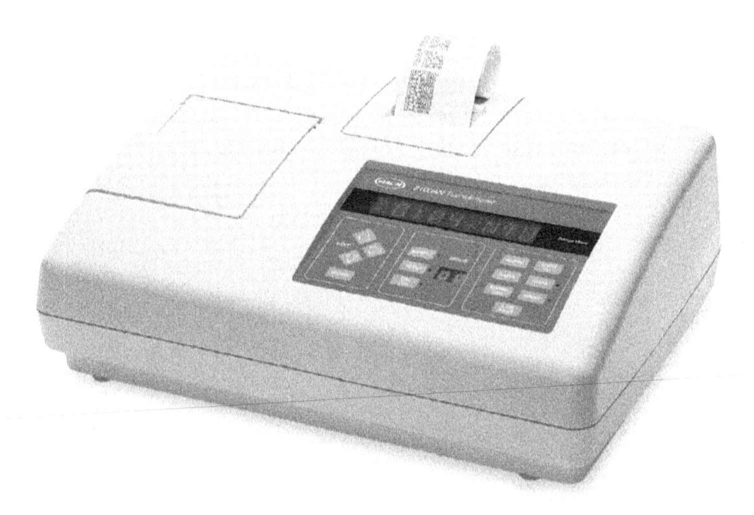

Source: Hach Company.

Figure 12-3 Bench-top turbidimeter. Unit is capable of turbidity measurements in a very low range.

Tips: For colder water temperatures, samples of raw water just prior to jar testing should be taken so that the operator will work with the coolest water possible. If the jar testing is being performed for TOC removals, be sure to wipe the mixer shafts with a clean cloth to remove any oils that may have seeped from the bearings of the mixer. If more than two jars are being tested, it is advisable that two operators perform the testing—this allows for the simultaneous application of chemicals to the jars. Use a clean syringe to take samples from the tops of the settled jars rather than using the attached hose. An excellent reference for jar test procedures is the Partnership for Safe Water manual, available from AWWA.

BIBLIOGRAPHY

APHA, AWWA, and WEF (American Public Health Association, American Water Works Association, and Water Environment Federation). 2012. *Standard Methods for the Examination of Water and Wastewater*, 22nd ed. Washington, D.C.: APHA.

AWWA. 2011. M37—*Operational Control of Coagulation and Filtration Processes.* 3rd ed. Denver, Colo.: American Water Works Association.

AWWA. 2010. Principles and Practices of Water Supply Operations—*Water Quality*, 4th ed. Denver, Colo.: American Water Works Association.

Owen, C.A. 2001. Tampa Bay Water. Internal memo to author. October.

Pizzi, N.G. 2007. *Pretreatment Field Guide.* Denver, Colo.: American Water Works Association.

US Environmental Protection Agency (USEPA). 1989. Drinking Water—National Primary Drinking Water Regulations—total coliforms (including fecal coliforms and *E. coli*). *Federal Register*, 54:124:27544.

Water Supply Operations: Jar Testing. DVD. 2010. Denver, Colo.: American Water Works Association.

Chapter 13

Instrumentation and Control Equipment

The design of a functional instrumentation and process control system for a water treatment plant begins by establishing a basic control philosophy. The designer needs to know the period of time that the plant will operate (continuous versus partial days), the degree of automation and process control desired, the type of signal to be transmitted, and whether the plant will be operated using local or remote control. Some knowledge of the types of alarms to be used and other telemetry requirements and computerization desired will also be necessary. A water treatment plant, with all of its subprocesses, is a specialized system. Knowledge of these subprocesses and their relation to each other leads the design engineer to take a comprehensive approach.

This chapter acquaints the operator with that approach, so that he or she can assist and comment appropriately during the design phase. Operator involvement during the design phase can be quite useful. An operator should realize that the design of the facility, which takes a few months, will affect operations for as long as the plant is in service. An operator's involvement at this critical stage may save effort and difficulty later on.

PROCESS CONTROL

The main purposes of process control are to produce a quality product, be alerted to process or equipment malfunction, and provide data acquisition that allows for plant evaluation and reporting. Data acquisition is also used to make decisions on process changes, which are necessary from time to time. Plant size, difficulty of treatment, and the capability of the operating staff impact the types of controls specified. Availability of qualified maintenance and electronic technicians, whether in-house or outsourced, will also impact the choices that are made.

Control Methods

Controls provided in the plant can be manual, semiautomatic, automatic, and supervisory. It is a good idea to choose the least complex methodology available that meets the control objectives. This will differ from plant to plant and allows for more on-site operator involvement, which is important in process control. There is a tendency for some design engineers to provide control systems that take the operator away from the process. This is done with good intentions:—the engineer desires to ease the operator's tasks—but this can lead to problems.

Consider this: the process engineer who designs the components of the water treatment plant (WTP) may never have operated one or may never have had to cope with the issues in the supervision of one—shift work, sick-time call-out procedures, employee issues that arise, power outages, and all of the day-to-day stumbling blocks that get in the way of well-intentioned and well-designed process control configurations. It is a virtual certainty that the WTP that is being designed will at some point have to be operated under manual mode and under stressful conditions. If an operator wants to be able to handle the plant at such times, it is a good idea for the operator to participate in the beginning of the conceptual design phase. The designer probably does not have operating experience and therefore cannot entirely meet the operators' needs.

An example is the complete automation of a filter backwash process. It has become standard design to allow for filter wash to take place based on predetermined criteria (head loss, turbidity breakthrough, etc.) with no operator involvement. All of the valve turnings, sequences, pumping, and other necessary functions will proceed without flaw, freeing the operator to complete other tasks. The designer views this as positive. However, in that instance, the operator may not view the backwash for "hot spots," hose down the walls and piping during the wash, sample the floc as it travels over the weirs, view the "rate of rise," or notice any of the many problems that can occur during the wash. That filter wash may come at an unpredictable time—or at a time that is forced because of any number of other process issues. An operator does not learn about backwash, or any process, without witnessing it *and controlling it.*

Supervisory Control and Data Acquisition

Control systems found in most water treatment plants are commonly referred to as SCADA (supervisory control and data acquisition) systems (Figure 13-1). The importance of SCADA is enhanced in part because of the regulatory compliance framework of the Interim Enhanced Surface Water Treatment Rule (IESWTR) and other similar requirements. Gathering, archiving, and analyzing huge amounts of raw data, which is required for compliance, require a more sophisticated data system such as SCADA.

SCADA systems are powerful and diverse, but it is important to choose these systems carefully based on the true benefit to the water plant. Poorly

Source: Cleveland, Ohio, Division of Water.

Figure 13-1 Operator checking SCADA control console. Real-time plant operating data are made available to the operator on demand.

designed systems can cause operational problems, and so-called state-of-the-art systems can be obsolete in just a few years, making it difficult to find replacement parts. A real consideration is the complexity of the treatment process. In general, water treatment plants are not as complex as industrial chemical plants, and the need for high-level SCADA may be questionable. However, this concept must be balanced with the size of some plants; very large facilities generate large amounts of data, which may be a justification for a more complex system.

Manual control of process components such as mixers, flocculators, and some valves is desirable. It is important to consider the restart of this equipment after a power failure. Continuous-operating equipment, like a flocculator, should be automatically restarted if the power fails. Operator safety, however, is an overriding criterion that must be considered whenever automatic restart is chosen. Semiautomatic controls are specified for processes such as initiation of backwash or other batch processing. Automatic and supervisory controls, which lend themselves to rate-of-flow control, chlorination, and sludge handling, must be carefully chosen.

Control Strategies

The two basic types of automatic controls are discrete control and continuous control. Discrete control ties equipment status (on–off) to status changes or programmed events. An example of discrete control is the automatic start–stop sequence for high-service pumps based on clearwell level. Continuous control requires an analog measurement for its input and manipulates a final control as an output. An example of continuous control is the feeding of chemicals based on fluctuating raw water flow. Chemical dosage is increased or decreased based on the flow rate signal.

Continuous control takes many forms, and they are discussed in the following sections.

Feedback Control

In the classic control-loop function of feedback control (Figure 13-2), a controlled variable is "fed back" to the controller for the purpose of detecting deviation from a set point. This system checks and rechecks actual system output against a desired value. In this way, it automatically corrects deviation, thus proving its value to the operator. The limitation of this system is that correction can only be made after deviation has occurred and elements such as meters, transmitters, and transducers are needed.

Feed-Forward Control

In feed-forward control, a process is controlled without the use of feedback—the control action is initiated before deviation occurs. Many water treatment plants use this type of control for simple chemical feed. The feeder output is controlled in proportion to plant flow, and the dosage goal is set manually at a ratio station. This simple controlling system can be used where accuracy is not critical and when time lags in the system will not take on great importance.

Two-Position Control

The simplest form of automatic feedback control is the two-position feedback system. It is called *on–off control* but is not limited to on–off situations. An example of its use is found in on–off pumping that is based on reservoir level. It is used when close control is not required.

Floating Control

Floating control is a three-position control that is used where control action is reversible. It is a variation of the on–off control but has the added elements of a second "on" position and a dead band. An example is the control of valves where a sequence of opening–off–closing is needed. This application is valuable because of its ability to accept operating times, such as the amount of time necessary to open a valve. It offers a tighter amount of control, or narrower range, than on–off control and will usually cycle to a lesser degree.

Floating Proportional Control

Floating proportional control is the on–off control equivalent of a conventional analog controller, with automatic reset features included (Figure 13-3). The advantages it has over analog systems are the simplicity of the on–off control and the fact that it will not lose control action when signal is lost.

Source: ICS Healy-Ruff.

Figure 13-2 Feedback monitor controller

This is a great advantage to the operator because it can eliminate sudden process upset.

Proportional Control

Simply stated, a proportional controller uses a mode where the output of the controller is directly proportional to the deviation from the set point. This type of system is used frequently for filter level control.

Proportional Plus Reset Control

The inherent limitation in proportional control is its inability to eliminate deviation from the set point over the control range. To overcome this, reset control is added. The reset function of this controller attempts to constantly change the controller output as long as there is deviation from the set point. When the process needs a wide proportional band for stable operation, this application is specified.

Proportional Plus Reset Plus Derivative Control

Proportional plus reset plus derivative control increases the response of the proportional plus reset control and is used only on slow-responding processes. These controllers are seldom used in water plants.

Accuracy

The degree to which an actual value will conform to an indicated value is the accuracy of the system. Accuracy, expressed as percent, is usually

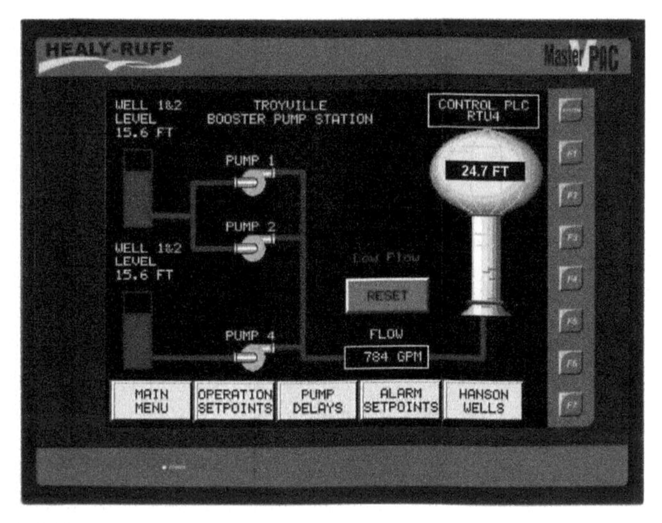

Source: ICS Healy-Ruff.

Figure 13-3 Pump controller

defined or specified as a percent of range, a percent of actual value, or a percent of range of a measured variable.

Accuracy and precision (repeatability) are not the same. Figure 13-4 shows the difference between the two. The figure depicts two targets that have been used by two sharpshooters. Sharpshooter A has placed five shots near the bull's-eye—one shot has found its mark. The other four are close, close enough in fact to consider sharpshooter A to be accurate. Sharpshooter A was not able to repeat the performance—he was not precise—but his performance may have been acceptable for the application intended.

Sharpshooter B was very precise in his pattern. He did not hit the target but all of his shots were placed in a tight pattern. His performance was repeatable but unacceptably inaccurate. He needs to adjust his rifle sight to find success.

Real-Time Measurement

Process control is accomplished using measurement or detection devices. Online instrumentation allows for continuous measurement of a process and therefore for immediate and accurate control of that process. Popular online measuring devices include pressure indicators, level indicators, online turbidimeters, chlorine residual recorders, and flow indicators. Offline instruments, such as laboratory devices, although not commonly used for controlling a process, are important to the operator. They help measure the effectiveness of a process for compliance purposes and also serve as calibration devices for online instrumentation.

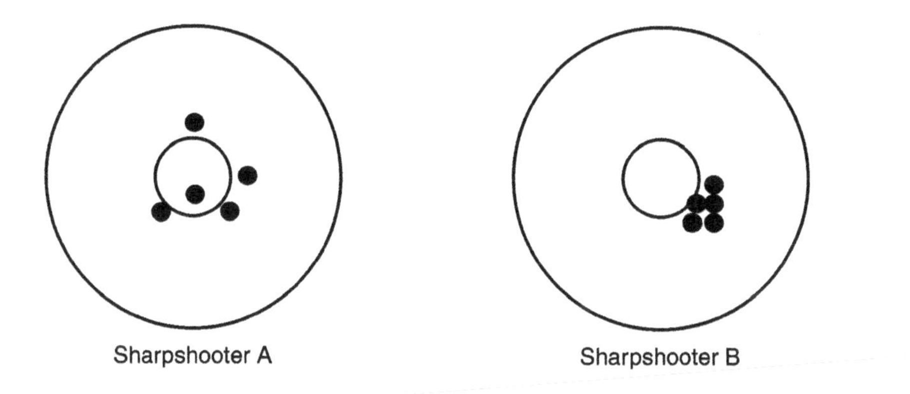

Figure 13-4 Schematic of the sharpshooter concept. Sharpshooter B is not accurate but is consistent. Sight adjustment is necessary.

Level

Operators need a reliable and accurate method of measuring levels in the plant, whether those levels are in the clearwell, the day tank, or some other location. Common methods for measuring and transmitting levels are float-type level indicators, bubbler units, pressure-actuated level transmitters, and sonic-type indicators. Each has advantages and disadvantages, and operators have their personal likes and dislikes. The simplest device is the float-type sensor. However, it is not practical where freezing will occur and it requires maintenance to provide a relatively friction-free motion for the cable assembly. Sonic devices, which work readily on a 4- to 20-mA system and require temperature correction, are popular.

Pressure

Gauges and transmitters for pressure sensing work in one of two ways. The first method uses a manometer, which balances the weight of a liquid, such as mercury, against the pressure to be measured. The second method measures the distortion of a metal caused by the pressure of water against that metal. Pressure is measured as absolute, gauge, or differential. Absolute pressure is referenced to an absolute vacuum, which is zero. Gauge pressure is measured with reference to the ambient atmospheric pressure, which is about 14.7 psi at sea level. Differential pressure shows readings that are the difference between two pressure readings.

Manometers use a U-shaped glass tube, which is etched with gradations of pressure. The tube is filled with mercury, which is 13.6 times as heavy as water. The pressure of the water pushes the mercury up the tube and a reading is taken. Due to the cost and hazards associated with mercury, manometers are being replaced with other types of pressure sensors.

Most pressure-sensing devices used in water treatment plants rely on the metal-distortion principle. Differential pressure cells also use this method. Water pressure is introduced to one side of a metal diaphragm, and the distortion is measured and translated as pressure, or pressure differential.

Flow

Flow is the most difficult and the most expensive physical measurement taken in the plant (Figures 13-5 and 13-6). Cheap measuring systems often result in high head loss measurements and inaccuracies; more expensive devices can be quite inflexible. The flow rate of water in a treatment plant is measured using various types of meters, including loss-of-head meters, positive-displacement meters, magnetic flowmeters, Venturi meters, rotameters, flumes, and weirs. For large flows, Venturis and flumes or weirs are most often chosen. They are nonmechanical devices that need little maintenance. Very low flow rates, such as for chlorinators, are usually measured by rotameters.

Filter Head Loss

Measuring loss of head in a filter is important because many operators determine filter run length based on this measurement (see chapter 7). Most plants have a filter strategy in place that requires filter run termination based on a preset loss of head, although the trend is to terminate based on filter effluent quality (nephelometric turbidity units [ntu], particle counts) because of public health considerations. Nevertheless, it is necessary to measure head loss in a filter. Head loss development that deviates from the norm, or historical, can provide the operator with valuable information.

Particulate Contamination

Particulate contamination in water is measured to determine the efficiency of the treatment process and as proof of compliance with regulations. As discussed in chapter 1, the Surface Water Treatment Rule and the IESWTR regulate the turbidity of the finished water. A goal of treatment is to reduce, as much as possible, the amount of source water particulate contamination before the water is delivered to the customer.

Turbidity

Turbidity is an optical property of water. When light is directed into a water sample, some light passes through the water, while some light is scattered. Some light is also absorbed by particles and dissolved substances in the water. Some turbidimeters and turbidity measurement methods provide correction for interferences from light-absorbing substances. Some do not. Water with high turbidity will scatter more light and appear cloudier than low-turbidity water.

Turbidity is measured in one of two ways: by measuring the light that is scattered or by measuring the light that passes through the suspension.

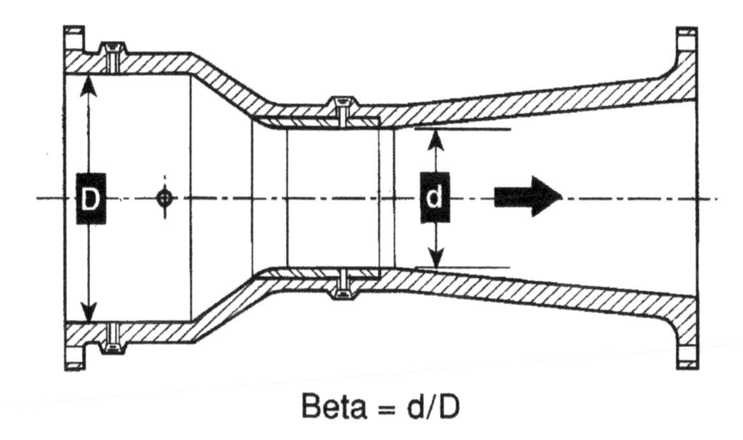

Source: Primary Flow Signal Inc.

Figure 13-5 Schematic of a flowmeter

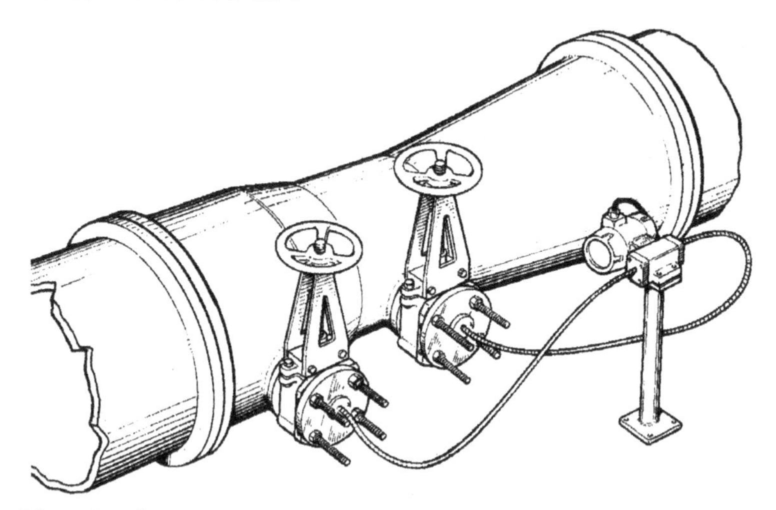

Source: Primary Flow Signal Inc.

Figure 13-6 Installation of a flow measurement device

The first method, called the *nephelometric method*, is commonly used in water treatment plants because it is sensitive and precise in clearer water. The second method is outdated and not commonly used in water treatment plants.

Turbidity measurements are taken with bench-top, portable, and online instruments. For regulatory reporting purposes, online and bench-top units are used. Online instruments are commonly used to monitor the effluent of individual filters, while bench-top instruments are often used to monitor raw, settled, and combined effluent, or "plant tap," water. As previously mentioned, the IESWTR requires that all filters in surface water treatment plants be monitored with online instrumentation.

Bench-top turbidimeters are batch-sample analyzers designed for grab samples, although some instruments can be equipped with flow-through cells that measure continuously. Bench-top units use glass cuvettes that hold approximately 30 mL of sample. The cleanliness of the cuvettes is critical for obtaining meaningful measurements. If multiple samples are taken and measured at one sitting, *indexing* (the reading contribution from an imperfections in the sample container, cuvette) must be used to minimize the deviation in readings brought about by the optical properties of the glass cuvette. New cuvettes should be tested for this deviation; any cuvettes that do not match should be discarded. Cells that deviate more than 0.01 ntu when measuring low-level turbidity should be rejected.

Cuvettes must be free of dirt, water spots, and fingerprints and should not be stored in an open container. Silicone oil is applied to the glass surface for cleaning and to minimize spots. Some operators use devices (sonicators) to degas water samples, i.e., remove air, which can contribute to false readings. Care must be taken when using these devices because they can cause dirt particles to break off of the cell walls of the cuvette.

Instruments should be calibrated according to manufacturer's instructions, which are based on constructing a calibration curve that covers the range of expected measurements. Calibration is accomplished using primary and secondary standards. Primary standards are prepared from traceable raw materials and include formazin, approved by USEPA, and as required by state regulatory agencies, for this purpose. Quarterly use of primary standards is usually required. Secondary standards are those that a manufacturer has certified to give instrument calibration within a certain tolerance when the instrument has been calibrated with the primary standard. Secondary standards, which are used daily for bench-top instruments, are very stable.

Online turbidimeters sample a side stream that is drawn from the treatment process. The water being measured flows from the process to the instrument's analyzer chamber, and measurements are made continuously. Online turbidimeters may be used at any point where a water supply can be tapped to provide representative, reliable sample flow and where the instrument itself can be accessed for routine maintenance and calibration.

Particle Count

Particle counting (Figure 13-7) provides the water plant operator with the potential to enhance the monitoring and control of water treatment processes. Although there is debate about the significance of particle counts and removal efficiencies as they correlate to pathogen removal, most utility operators who use particle-counting devices believe that they enhance plant operation. Correlation between particle counts and turbidity in water varies depending on the cause of the particles or turbidity. Specifically, turbidity fluctuates more as a result of the smallest particles in water (less than 1 µm in size, such as colloids commonly found in treated water) and particle

Source: Joe Zimmerman, Chemtrac Inc.

Figure 13-7 Particle counter

measurements fluctuate more due to larger particles such as coagulated solids breaking through or off filters or formed postfiltration via chemical precipitation.

The operator who effectively uses both online turbidimeters and particle counters will come to better understand the total particle characteristics of filtered water. For example, through the course of a filter run, smaller particles (as measured by turbidity) will continually decrease; near the end of the run, larger particles (as measured by particle counters) will begin to detach. This detection of elevated particles prior to turbidity breakthrough allows the operator to further optimize filter washing.

A goal for filter effluent production is set by choosing total particle count. For example, an operator might measure particle counts greater than 2 µm in size and attempt to keep those counts below 50/mL in the filter effluent. When the filter effluent's particle count is greater than 50/mL, the filter is taken out of service and backwash variations are implemented in treatment chemistry or filter backwash procedures. It has been advocated that operators track and respond to increases in particles in specified size ranges, such as those that are 3 to 5 µm in size, as a possible surrogate for *Cryptosporidium*. Because particles of all sizes have been shown to be removed with similar methods, it is usually best practice for operators to track and respond to total particle counts (>2 or 3 µm in size, depending on the lower detection limit of the counter). This gives the largest counts to work with and hence provides the greatest potential for noticing and responding to any elevated counts.

Particle counters operate by light-blocking, or by light-scattering, principles. Those commonly used in drinking water analyses may provide different counts due to their specific characteristics, e.g., calibration or make. There is little agreement at this time as to calibration standards. For this reason, particle counts are not regulated, as is turbidity, but particle counters can be a very useful tool to compliment turbidimeters for use in treatment optimization.

Temperature

Temperature can affect process control systems, laboratory analysis, and equipment operation. The measurement of temperature is a component of regulatory compliance when calculating $C \times T$ (see chapter 8). Temperature is commonly measured using thermocouples, thermometers, and bimetallic units. Thermocouples operate on the principle that an electromotive force is generated when heat is applied to one of the junctions of two dissimilar metals. The electromotive force, which is proportional to the difference in temperature at the two junctions, is measured and displayed as temperature.

The thermometer, or thermal bulb, operates on the principle that the absolute pressure of a gas is proportional to the absolute temperature. Bimetallic units are constructed of two different metals bonded together. Each metal expands and bends at a different rate and thus responds differently to different temperatures. The amount of bend is measured and converted to temperature.

Chemical Measurements

pH

The term pH is defined as the negative logarithm of hydrogen ion activity, and it covers a scale of values from 0 to 14. The measurement of pH requires an electrode, which is comprised of a pH-sensitive device, a reference electrode, and a temperature-compensating mechanism. The electrical potential created by hydrogen ion activity is measured directly by the pH-sensitive device; the reference electrode measures the ambient electrical potential; and the temperature cell compensates for any influences of temperature. Although pH measurement is a frequent bench-top laboratory function, operators also use pH probes directly in a process to monitor and control efficiency, especially in softening plants.

Chlorine Residual

Chlorine residual analyzers and recorders, both of which are very important to the operator, play a dual role in water treatment. As controlling devices, they use a closed-loop feedback system to control chlorine dose in close tolerances. As recorders, the use of which is regulated, they provide a record of the chlorine residual leaving the plant and are designed to notify the operator of pending problems with low or high amounts of chlorine.

Two electrochemical principles are used in chlorine residual controllers: amperometric and polarographic. An amperometric device uses a reagent to maintain a constant pH and an electrode to measure electrical current that is proportional to the chlorine residual. Free residual or total residual is measured using buffer solutions. Electrodes are susceptible to interferences from manganese, iron, nitrite, and chromate. A polarographic sensor contains a chlorine gas–permeable membrane. As chlorine traverses the membrane, it causes an electrical current that is proportional to the chlorine residual to be generated. This current is measured and displayed. Operators need to have ready access to this equipment while making rounds.

Analyzers that use colorimetric methods (DPD) are also available. The systems use standardized color indicators and the intensity of the color to determine the chlorine concentration.

Chlorine residual analyzers and recorders are often used to control chlorine dosage based on the residual measure. Controlling mechanisms are pictured in Figures 13-8 and 13-9.

Total Organic Carbon

Organic compounds in water have taken on great importance due to regulation of disinfection by-products. Consequently, it is desirable to measure the overall amount of organic carbon in water. The TOC measurement was developed for this purpose. Several continuous-flow online TOC instruments are available. Most units depend on the extraction of inorganic

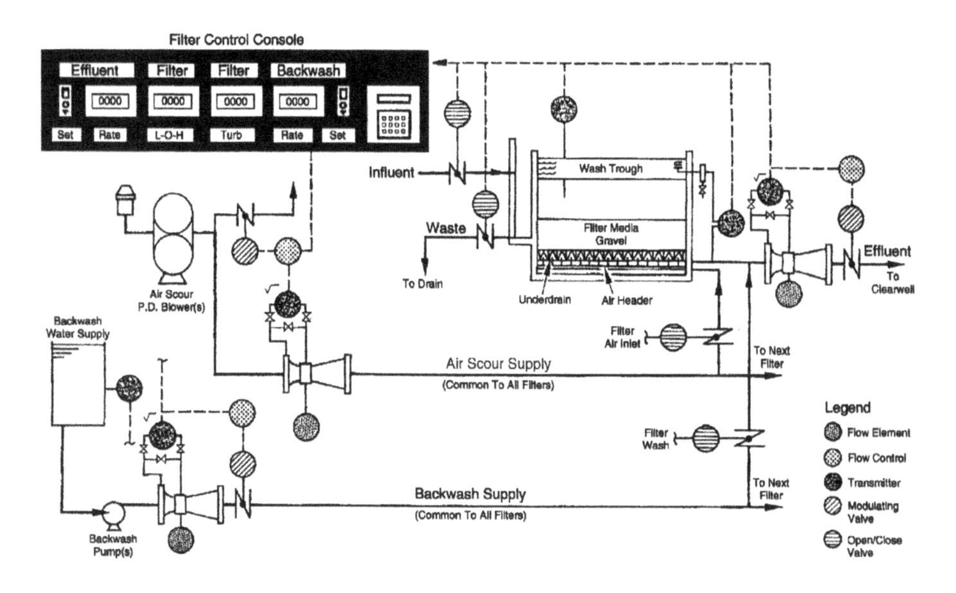

Source: F.B. Leopold, Co. Inc., Zelienople, PA 16063.

Figure 13-8 Filter control schematic

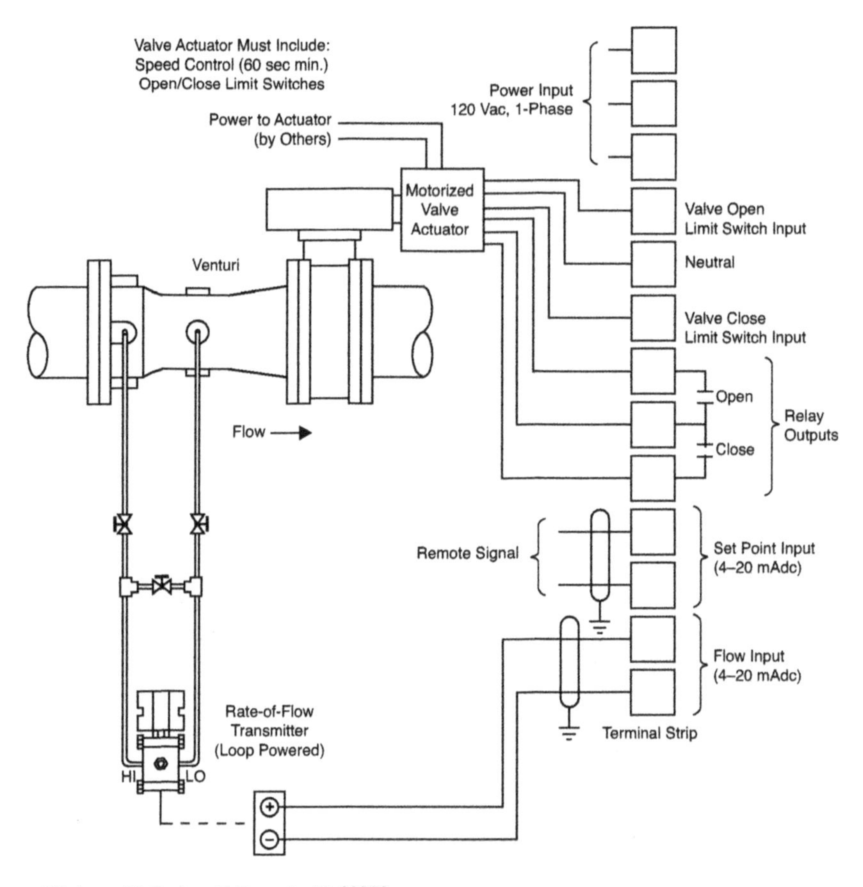

Source: F.B. Leopold, Co. Inc., Zelienople, PA 16063.

Figure 13-9 Schematic of a valve positioner

carbon, which is then converted to carbon dioxide and combusted. The carbon that remains after combustion is organic.

There are several major types of instrument technologies used for online TOC analyzers including:

- High temperature catalytic/combustion oxidation (HTCO)
- Supercritical water oxidation (SCWO)
- Patented two-stage advanced oxidation (TSAO) process using hydroxyl radicals
- Heated persulfate ultraviolet (UV) oxidation

Other Online Chemical Analyzers

Online analyzers have been developed for many other chemicals in water. The accuracy and reliability of each must be evaluated for the intended purpose. Also, some of these analyzer test methods are not approved for compliance testing. Before purchasing an online analyzer, the operator should pilot test the unit being considered under real plant conditions.

Iron, manganese, and phosphates are among the chemicals that are most often tested using online continuous monitors. The test methods are usually either colorimetric or a type of selective electrode. The units must be calibrated and operated according to manufacturer's instructions to give the best results.

BIBLIOGRAPHY

AWWA. 2001. M2—*Instrumentation and Control*, 3rd ed. Denver, Colo.: American Water Works Association.

Burlingame, G.A., M.J. Pickel, and J.T. Roman. 1998. Practical Applications of Turbidity Monitoring. *Jour. AWWA*, 90(8):57–69.

Guidelines for Design, Construction, and Operation of Water and Sewerage Systems, Section 6, Instrumentation and Control. Government of Newfoundland and Labrador, Department of Environment and Conservation, Water Resources Management Division. December 2005.

Letterman, R.D., S. Viswanathan, and J. Dwarakanathan. 2001. *A Study of Low-Level Turbidity Measurements.* Denver, Colo.: American Water Works Association and American Water Works Association Research Foundation.

McTigue, N.E., M. LeChevallier, H. Arora, and J. Clancy. 1998. *National Assessment of Particle Removal by Filtration.* Denver, Colo.: American Water Works Association and AwwaRF.

MWH, Consulting Engineers Inc. 2012. *Water Treatment:Principles and Design.* Third ed., New York: John Wiley and Sons.

Recommended Standards for Water Works. 2012 Ed. A Report of the Water Supply Committee of the Great Lakes–Upper Mississippi River Board of State and Provincial Public Health and Environmental Managers, Health Research Inc., Health Education Services Division. http://www.healthresearch.org

Routt, J.C. 2001. Personal communication. October.

Chapter 14

Safety and Security Practices

TREATMENT PLANT SAFETY

The discussion of plant safety presented in this chapter is not intended to be a comprehensive guide on the subject. It is the responsibility of utility personnel to follow safety regulations and to be properly trained to perform their duties safely. The bibliography at the end of this chapter provides additional information on safe working practices. The information provided in this chapter should be adequate for most operator certification test examinations.

Safety and safe working conditions are the responsibility of each individual associated with a water utility. The degree of responsibility varies among upper management, supervisors, and operations personnel, but all play a role and all are responsible for their individual actions. Safety regulations guide those actions, but the creation of policies, procedures, and methods to meet safety requirements is the result of a collaborative effort within the water utility.

Utility managers are responsible for maintaining safe working conditions, the physical condition of their facilities, and supporting policies that encourage the safe performance of work duties. Supervisors at all levels directly control all work conditions and are responsible for the activities of the personnel they supervise. Each supervisor, foreman, crew chief, or lead staffer is responsible for ensuring that all work is done in compliance with safety practices, utility policies, and regulations.

Employees have a particular responsibility when it comes to safety. That responsibility is to correctly use the safety equipment provided and follow all safety policies and procedures. All employees must be aware of the safety requirements and help guard against unsafe acts and conditions.

SAFETY REGULATIONS

The primary reason for developing and maintaining safe working conditions and practices is to eliminate injuries. Beyond the personal cost to an injured employee are the costs to the utility from lost time, medical expenses, and possible legal judgments. Other considerations include damage to equipment and property and resulting repair costs, and the potential need to hire and train new employees to perform the work duties of the injured employee.

Occupational Safety and Health Administration Act

Another reason for developing safety policies and procedures is the federal Occupational Safety and Health Administration (OSHA) Act (Public Law 91-596). Passed in 1971, the act established OSHA and compiled numerous safety and health standards that are applicable to every industry. Specific standards have been developed for most work activities and chemical substances that an employee may be exposed to in the course of his or her employment. These are minimum standards that must be followed; their requirements are itemized in the Code of Federal Regulations (CFR) (29 CFR, 1910 and 1926). The act provides for monetary penalties, which are escalated depending on the seriousness of the safety violation, and the potential for incarceration if OSHA safety standards are violated.

If a specific requirement does not exist, OSHA relies on the general duty clause, which requires each employer to provide employment, and a place of employment, that is free of recognized hazards that are likely to cause death or serious physical harm. The general duty clause also requires employees to comply with safety practices and procedures. In addition to OSHA, many states and cities have their own safety requirements. In most cases, these standards mirror the OSHA standards; however, in some cases, state or local requirements are more stringent and carry additional penalties for failing to comply.

CONFINED SPACE RULES

A common hazard encountered in the water industry is confined spaces. Examples of confined spaces in the water and wastewater industry include access holes for valves, meters, and air vents; chemical storage tanks or hoppers; wet wells; digesters; sedimentation basins; filters; and reservoirs.

One of the most sobering statistics related to confined space safety is one relating to the death rate for those entering confined spaces. Although the death rate relating to confined spaces continues to decrease, the percentage of those who die while attempting a confined-space rescue remains roughly the same. Each year about two-thirds of those who die in confined spaces are would-be rescuers. The cause of death is usually asphyxiation or the result of an atmospheric hazard that could not be seen.

Under OSHA regulations, there are two basic types of confined spaces—permit required and nonpermit required. For a location to be classified as a confined space, the following three criteria must be met: the space must have limited means of entry and exit, the space must not be designed for continuous human occupancy, and the space must be of a size and configuration that humans can enter the space to perform work. If all three criteria are met, it is necessary to assess the space and determine if the space is also permit-required. A permit-required confined space meets one or more of the following criteria:

- The space contains, or has the potential to contain, a hazardous atmosphere
- The material within the space has the potential to engulf the entrant
- The internal configuration is such that it could trap the entrant (i.e., downward sloping and converging floors)
- The space contains any other recognized serious health or safety hazard

If the only hazard is atmospheric and the hazard can be controlled with ventilation, the space may be reclassified as nonpermit required. Figure 14-1 is a decision tree that can be used to identify the need for permitting a confined space.

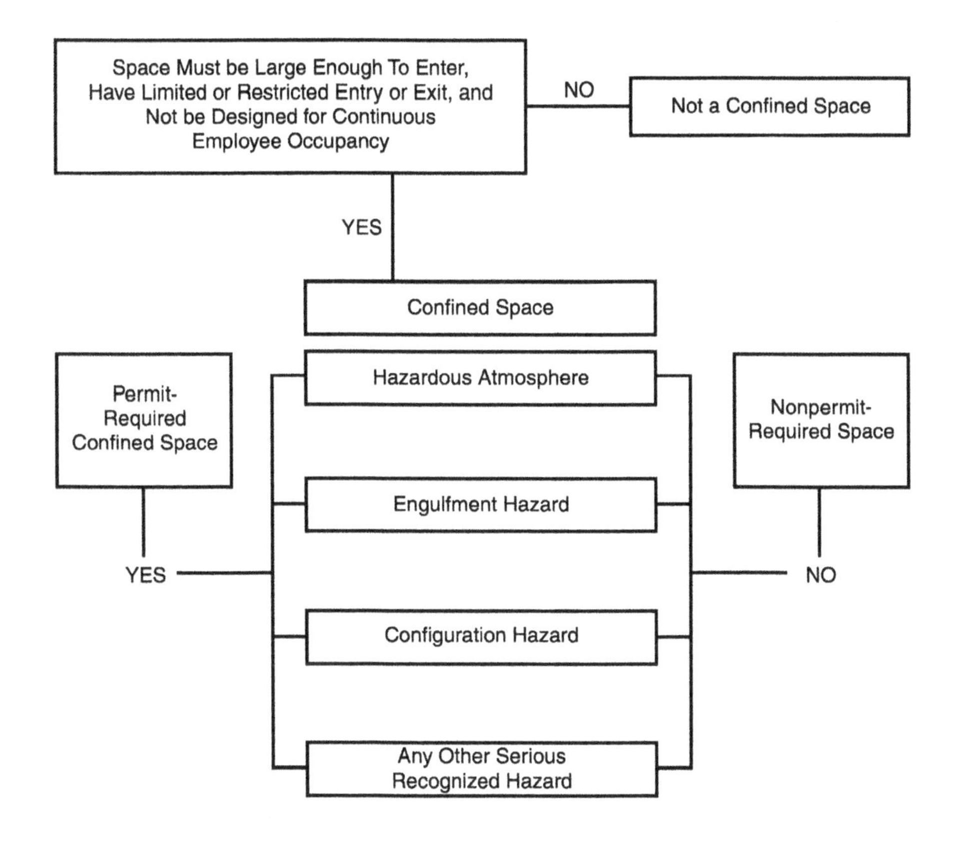

Source: Melinda Raimann, Cleveland, Ohio, Division of Water.

Figure 14-1 Confined space decision tree

A written program is required when it is necessary to enter permit-required confined spaces. This program must: contain a mechanism for identifying and controlling the hazard, have a written entry permit, provide for the identification and labeling of confined spaces, and provide employee training, among other requirements. A key component of this program is the requirement that before an employee enters a permit-required confined space, the internal atmosphere of the space be tested with a calibrated, direct-reading instrument capable of measuring oxygen content, presence of flammable gases and vapors, and potentially toxic air contaminants. In addition to these pre-entry tests, the atmosphere must be retested while work is being performed inside the space to ensure that acceptable conditions are being maintained. More details regarding the required program can be found in 29 CFR 1910.146.

An employer is required to provide its employees with the specialized equipment that is required for confined space entry. This equipment includes the following:

- Testing and monitoring equipment
- Ventilation equipment (to eliminate or control atmospheric hazards)
- Communications equipment, as necessary
- Personal protective equipment (PPE)
- Lighting equipment, as necessary
- Barriers and shields to prevent unauthorized entry, as necessary
- Ladders for safe entry and exit
- Fall-prevention equipment required because of the difference in elevation
- Rescue and emergency equipment, including harnesses and hoists, self-contained breathing apparatus (SCBA), stretchers or backboards, and supplies

There are three classes of employees, as they relate to confined spaces: the entry supervisor, the authorized entrant, and the authorized attendant.

The *entry supervisor* is responsible for knowing the conditions within the confined space, verifying that all equipment and procedures are in place prior to entry, verifying the availability of rescue services and the means of summoning them, terminating entries and canceling permits, and determining that acceptable conditions as specified in the permit continue for the duration of the entry.

The *authorized entrant* is responsible for knowing what hazards are to be faced, recognizing the signs and symptoms of exposure and understanding the potential consequences, knowing how to use any needed equipment, communicating with the attendant as necessary, alerting the attendant when a warning symptom or other hazardous condition exists, and quickly exiting when ordered or alerted.

Authorized attendants are responsible for being aware of the confined space hazards and the behavioral effects of exposure; maintaining a count and the identity of the authorized entrants; preventing unauthorized entry;

remaining outside the space until relieved; communicating with the entrants and monitoring activities inside and outside of the permit space; and ordering exit, summoning rescuers, and performing nonentry rescue as required.

GENERAL PLANT SAFETY

Right to Know and MSDS

In order to follow safety practices and procedures that will protect an employee and fellow workers, an employee must be aware of the hazards to which he or she is exposed. A primary method of doing this is through a Right-to-Know program developed by the employer. The main premise of a Right-to-Know program is that an employee has the right to be advised of the hazards associated with the chemicals and materials that they work with or that exist nearby in the course of their daily job activities.

Chemicals, as defined under the Hazard Communications standard (29 CFR 1910.1200), include those used for water treatment, lab procedures, cleaning, and maintenance applications. A material safety data sheet (MSDS) (see appendix A) lists a chemical's identity, composition, and type or types; the hazard(s) associated with its use (is it flammable, an oxidizer, poisonous); how it enters the body; the effects of exposure; health effect (short term, long term); the permissible exposure limit (PEL); how to handle spills or releases; and other related information. An MSDS for every chemical that is used in a water treatment plant should be kept in an organized fashion in a location that is readily accessible to all employees.

Employees are to be trained on the hazards of the chemicals they use or are exposed to and in the methods used to protect themselves from those hazards.

Risk Management and Emergency Response

The US Environmental Protection Agency (USEPA) issued the Risk Management Rule on July 20, 1996, in an effort to protect public health and safety. OSHA had previously released the Process Safety Rule in February 1992 (29 CFR 1910.119), which focused on employee health and safety but did not address the safety of nonemployees. According to the Risk Management Rule, employers must not store or use chemicals in quantities greater than the threshold quantity, which is defined by USEPA (e.g., more than 1,500 lb of chlorine), and must develop release scenarios, conduct hazard assessments, determine means and methods to limit public exposure should a spill or release occur, and provide the public with information about their plan.

Emergency response at a water treatment facility generally means responding to a chemical release. This response is governed under the Risk Management Rule and under OSHA (29 CFR 1910.1). A number of decisions must be made regarding emergency response, the first of which is to determine who will respond in the event of a chemical release. If the utility elects to have its employees respond to chemical releases, the utility is required to comply with the requirements of OSHA's Hazardous Waste Operations and Emergency Response (HAZWOPER) regulations.

Compliance requires the employer to provide specialized chemical response suits and respirators, medical exams to determine the fitness of the employees designated as responders, spill-containment materials, and annual training. Initial training is usually 24 to 40 hours in duration, depending on the responsibilities the response team is expected to assume; an annual 8-hour refresher course is also required. Even if a utility chooses to have others act as the designated emergency response team, the HAZWOPER regulations, as they relate to awareness-level training for those who may be exposed during a chemical release, are to be followed. This awareness-level training is an eight-hour course with an annual refresher that focuses on the hazards of the chemicals used and evacuation procedures, routes, and staging areas to be used.

A key component of any emergency response plan is for the responder to be aware of the hazards of the chemicals that may be released and what precautions must be taken to ensure the safety and well-being of all designated responders. This information is usually provided on the specific MSDS or as part of the facility's risk management or hazard communications program.

Standard practice for emergency response is to follow the incident response system first developed by fire and safety forces and designate someone in charge of the emergency site. Red (hot), yellow (decontamination), and green (clean) zones should be designated, and strict controls are used to ensure that no unauthorized or unprotected entrant enters the chemical-release area. A spill or release area should never be entered alone; one standby person, fully suited in the appropriate level of PPE, should be assigned to each entrant into the hot zone.

First Aid

Water treatment plants often provide first-aid stations for their employees. These stations are stocked with emergency supplies that are used when operators sustain minor injuries on the job. The stations (Figure 14-2) may include oxygen, bandages and antiseptics, fire blankets, and eyewash solutions. SCBA equipment may also be stored here if the location is near the chlorine area. Be sure to document use of all supplies immediately after use.

Chlorine Safety

Chlorine safety is a subset of chemical safety at many water treatment plants. Chlorine is a toxic chemical and is an irritant to the eyes, skin, mucous membranes, and respiratory system. Effects of exposure generally are evident first in the respiratory system and then in the eyes. The impact of exposure to chlorine is dependent on both concentration and time. The very young, the elderly, and people with respiratory problems are most susceptible to chlorine's effects. As the duration of exposure or the concentration increases,

Source: Conneaut, Ohio, Water Department.

Figure 14-2 First-aid and safety station

the affected individual may become apprehensive and restless, with coughing accompanied by throat irritation, sneezing, and excess salivation. At higher levels, vomiting associated with labored breathing can occur. In extreme cases, difficulty in breathing can progress to the point of death through suffocation. An exposed person with a preexisting medical or cardiovascular condition can have an exaggerated response.

Anyone exhibiting these symptoms should see a qualified health care provider immediately as his or her condition will deteriorate over the next few hours. The physiological effects from exposure to various levels of gaseous chlorine are presented in Table 14-1.

To protect employees who work with or around chlorine, the employer must ensure that chorine equipment is kept in safe working order, regular maintenance is performed on all chlorine equipment and the equipment used to handle it such as overhead cranes, and employees have the proper tools and PPE.

Chlorine cylinders of 150 lb are equipped with fusible plugs that will melt at temperatures between 157°F and 162°F and release the contents of the cylinder (Figure 14-3). Consequently, cylinders must be stored in a temperature-controlled area. Labels identifying the contents of any cylinder or tank car should be legible and prominently displayed. Placards with Department of Transportation (DOT) codes identifying the contents and their hazards are to be posted on the exterior of chlorine rooms so that emergency response teams can identify and address the hazards they will be exposed to in the event of a release. All cylinders should be secured and protected from accidental contact by moving equipment or vehicles.

Table 14-1 Physiological effects of chlorine exposure

Exposure Level, *ppm*	Effects
0.02–0.2	Odor threshold (varies by individual)
<0.5	No known acute or chronic effect
0.5	ACGIH (American Conference of Governmental Industrial Hygienists Inc.) 8-hour time-weighted average
1.0	OSHA ceiling level
1–10	Irritation of the eyes and mucous membranes of the upper respiratory tract. Severity of symptoms depends on concentrations and length of exposure.
3	ERPG-2 (Emergency Response Planning Guidelines as values developed by the American Industrial Hygiene Association) is the maximum airborne concentration below which it is believed that nearly all individuals could be exposed for up to 1 hour without experiencing or developing irreversible or other serious health effects that could impair an individual's ability to take protective action.
10	Immediately dangerous to life and health (per the National Institute of Occupational Safety and Health)
20	ERPG-3 is the maximum airborne concentration below which it is believed that nearly all individuals could be exposed for up to one hour without experiencing or developing life-threatening health effects.

Source: The Chlorine Institute.

Personal Protective Equipment

OSHA requirements governing PPE are found in 29 CFR 1910 subpart I (see appendix E in the CFR), which also references various American National Standards Institute guidelines. As a general rule, PPE is to be considered as a last resort. The elimination of the hazard through engineering changes is the optimal solution, followed by administrative changes. If these methods do not eliminate or minimize the hazard to the appropriate level, PPE is to be used. The PPE must be appropriate for the hazard it is being provided to protect the employee against. This can only be determined through a hazard assessment and equipment evaluation. In general, PPE is provided by the employer, unless it is used voluntarily or by an employee outside the work environment. In addition to providing the equipment, the employer is also responsible for ensuring that it is maintained in a sanitary and reliable condition.

PPE can be provided for nearly every portion of the body. Occupational foot protection, or safety shoes, should be used when there is the potential for injury to the foot from falling, rolling, or piercing objects or when extreme thermal conditions exist. Different types of shoes or devices provide different types of foot protection. There are steel toe caps, both internal and external to the shoe; metatarsal guards; steel shanks; and various combinations of these devices. They are commonly worn in machine shops and during construction activities.

Source: Conneaut, Ohio, Water Department.

Figure 14-3 Safely stored 150-lb chlorine cylinders. Caps are on and the chain is set.

Another common piece of PPE is the hard hat, which is required when there is the potential for injury to the head from impact or falling objects. In a water treatment plant, hard hats are usually worn when an overhead crane is in use or during construction projects. They may be required if there are low vertical clearances and an employee has to crouch or duck to travel through the area, such as in a pipe gallery or subbasement location. There are also specific types of hard hats designed to provide protection from high-voltage shocks and burns.

Treatment plants typically have high noise areas, usually on pump floors or in machine shops. If the employer is unable to lower the noise levels or limit employee exposure, hearing protection must be provided. Hearing protection comes in various forms, from earplugs to muffs, and offers varying degrees of protection. Noise-level sampling is required to determine what level of noise reduction is needed, which will be marked on the protective devices as a noise reduction reading. As a general rule of thumb, if an employee must raise his or her voice to be heard in the area in question, hearing protection is probably required. Exposures to noise levels above an

8-hour time-weighted average of 85 dB_A (decibels measured on the A scale) require a hearing-conservation program, which includes monitoring. Specific requirements are found in 29 CFR 1910, subpart G.

Eye and face protection may also be required when handling chemicals or doing work that produces flying debris. Impact-resistant eye and face protection (glasses, goggles, face shields) are required if flying debris is likely to be present. Chemical splash–resistant and/or vapor-resistant equipment is required for chemical handling, as specified on the chemical MSDS. Body protection, such as aprons or coats, may also be required for handling chemicals. Again, refer to the MSDS for the appropriate type and composition of protective device to be selected. The employer is responsible for providing the equipment and ensuring that it is maintained. The employee is responsible for using it as required.

Self-Contained Breathing Apparatus

Self-contained breathing apparatus (SCBA) are a type of respirator that supplies safe, grade D (or better) breathing air to the wearer. They are commonly used in the water and wastewater industries for emergency response or for employee protection when maintenance activities can reasonably be expected to result in the release of a hazardous material. They are also required in oxygen-deficient atmospheres or where the existing atmosphere is immediately dangerous to life and health. The other type of respirator that supplies breathing air is an airline respirator. They are not commonly used for emergency response because the length of the airline supplying the breathing air is only 300 ft. This type of respirator will not be discussed in this section. The major components of an SCBA include an air tank, harness for wearing the tank on the back, pressure regulator, hoses, face mask, and a low-air alarm; all components require routine inspection and maintenance. Refer to the manufacturers' literature for specific maintenance intervals. DOT also has testing, labeling, and maintenance requirements for air cylinders. OSHA inspection requirements mandate inspection during routine cleaning if respirators are worn routinely and at least monthly and after each use when used for emergency purposes.

OSHA requires employees to be medically fit to wear an SCBA. Wearing an SCBA puts additional strain on the human body; therefore, it is important to ensure that there are no underlying medical conditions that would limit an employee's ability to use the device. An employee must be medically evaluated before using an SCBA and on a periodic basis thereafter. Annual medical exams are no longer required, but annual review of an employee's fitness by a medical provider should be performed.

In addition to medical surveillance, OSHA requires fit testing of tight-fitting respirators. The two types of fit testing are qualitative and quantitative. Qualitative fit testing involves determining whether or not the employee can detect the odor of a test medium. Irritant smoke and banana oil are commonly used for this type of test. *Qualitative* fit testing is subject to the impression of the wearer and may not always yield an accurate result. *Quantitative* fit testing involves using a probed face mask identical in brand and size to the one assigned to the employee. The testing device measures the particles in the air outside the mask and the particles inside the mask and compares the levels to determine an acceptable fit. A number of manufacturers produce these testing devices. If an appropriate fit cannot be obtained with the assigned mask, another brand or size of respirator face mask may be needed to provide the required level of protection. Specific requirements regarding fit testing can be found in 29 CFR 1910.1050 (see appendix E in the CFR).

LAB SAFETY

General safety rules for laboratory workers or those who use chemical reagents in their duties can be found in the latest addition of *Standard Methods for the Examination of Water and Wastewater.* All water treatment plant employees should be familiar with the general rules. The previous discussion on PPE applies to laboratory workers in particular.

GENERAL SAFETY

Water plant structures can present opportunities to implement safety practices, and operators are urged to use a commonsense approach to identify these opportunities. Operators should work with designers of their treatment plant structures (Figures 14-4 and 14-5) to ensure safe accommodations are made a part of the overall safety plan.

PLANT SECURITY

Water Utility Security Initiatives

The Department of Homeland Security has taken steps in partnership with all public and private stakeholders to ensure the protection and resilience of water services. These Water Sector partners collaborate to be better prepared to prevent, detect, respond to, and recover from terrorist attacks and other intentional acts, and natural disasters, otherwise known as the *all-hazards* approach.

Homeland Security Presidential Directive (HSPD)-7 identifies 18 critical infrastructure and key resources sectors and assigns an agency lead for each. The USEPA is the designated agency for the water sector.

All-Hazards Approach

Water utilities have taken steps to protect critical water supply facilities from theft and sabotage, and have planned for response and recovery from events that interrupt water service. An all-hazards approach to security and emergency preparedness mirrors the time tested multi-barrier approach used in water treatment.

Source: Conneaut, Ohio, Water Department.

Figure 14-4 Guardrails at the 3-mgd Conneaut Water Treatment Plant. Ramp leads to chlorine storage area; 150-lb cylinders can be wheeled up the ramp.

Source: Sangre de Cristo Water Division.

Figure 14-5 Safety ladder arrangement at Santa Fe's lime silo. Note that tank climbing requires the use of a safety belt, which can be clamped to the railing as the climber ascends and descends.

The Water Sector Coordinating Council consists of representatives from a cross-section of water agencies and professionals. The council interacts with federal agencies regarding homeland security issues. In this role, the council has developed several critical resources for water utilities.

- *Water Sector–Specific Plan: Annex to National Infrastructure Protection Plan*
- *All-Hazard Consequence Management Planning for the Water Sector*
- *Roadmap to Secure Control Systems in the Water Sector*
- Water Sector Decontamination Priorities: Recommendation and Proposed Strategic Plan

In addition, standards have been developed as part of the AWWA/ANSI process to assist water systems in their efforts to improve facility security and emergency preparedness.

- G430-09 Security Practices for Operations and Management
- G440-11 Emergency Preparedness Practices
- J100-10 RAMCAP Standard for Risk and Resilience Management of Water and Wastewater Systems

Water plant operators should be aware of the requirements for the storage and transportation of hazardous substances. Specifically, chlorine gas and large quantities of sodium hydroxide (caustic soda) have extensive requirements for emergency response and security. Many treatment plants have elected to abandon using chlorine gas due to these requirements and concerns. On-site generation of sodium hypochlorite has gained popularity as a way to avoid these hazards.

Treatment Plant Security Measures

Protecting a public utility's assets to ensure an uninterrupted, safe water supply is a fundamental concern of all levels of management. Plant operating staff should participate in the development of a master plan that will establish a level of protection for their areas of responsibility. A utility can implement a fully developed master plan to provide for overall security for the utility. These plans are site specific but contain common components. By breaking down these components into areas of responsibility, a utility can assure an integrated approach to plant security development. The major features of a security master plan are described in the following paragraphs.

Physical Requirements

The physical requirements of the plan focus on actions that terrorists, saboteurs, or criminals might take to damage or contaminate the water supply or sabotage system infrastructure and the electric transmission system. Preventing acts of sabotage at water treatment plants, electric substations, maintenance facilities, other utility facilities, and in the distribution system requires physical barriers to prevent intruders from entering a site and reaching vital areas or equipment (Figure 14-6).

Integrated Systems Approach

An integrated systems approach that combines people, equipment, and procedures is used to protect against possible threats and intrusions. The

Figure 14-6 Decorative security fence

security plan has multiple layers of protection that provide several ways to monitor sites and provides deterrence measures for a multitude of possible threats. It also monitors employee activities and documents the presence of contractors, visitors, and other invitees to the facilities through the use of swipe cards and proximity card readers.

Detection Systems

Electronic detection systems, which are more reliable and predicable than systems that depend on people, provide alarms to alert security and plant personnel that the plant or facility is being approached or an intrusion attempt is under way. Many cost-effective devices are available to detect movements, forced entry, and even record any events and times. Video surveillance is another useful detection system that can aid operators to prevent unlawful entry to the plant facility.

Delaying Tactics

Barriers, such as chain-link fences around perimeters, gates, and proximity card readers, are arranged in layers to create physical and psychological deterrents and to delay intrusion long enough for security to detect the intrusion and respond. Barriers channel entry and exit through specified control points, facilitating efficient identification and control of authorized people and equipment.

Assessment Systems

Closed-circuit television cameras monitor facility entrances and special secured areas and allow security to assess an alarm received while limiting exposure until assistance arrives. These systems can be linked with other security systems to control entry and record images for use by authorities.

Communications and Response Systems

Security assessment and monitoring devices are linked to communications systems to transmit information to a local monitoring site, a central monitoring site, and the security force. As the security officer or facility personnel respond, they remain in communication with a central monitoring site and other site personnel at all times by several means of alternate communication, such as 800-Mhz radios, cell phones, walkie-talkies, pagers, and other communication devices available at the site.

Training and Qualifications

Security and facility personnel effectiveness must be maintained through training. Training exercises against mock adversaries should be conducted periodically to add realism to the training. In addition, training on system maintenance and reporting and policy enforcement must be conducted annually for effective system management.

Access Control

Employees are issued identification badges that must be carried at all times. Access control systems at the entrance gate and within buildings will confirm a person's identity and authorization before entrance is granted. Visitors are also issued badges upon entry to the facility.

Reliability of Personnel

Security awareness will become part of each employee's initiation to the work site. Thorough background investigations should be conducted in accordance with department policies as the primary method for learning about any past actions that may indicate problems with individual employees.

Testing, Maintenance, and Auditing

Periodic testing of all security equipment is required to make sure it is operating correctly. If problems are found, corrective measures are required to ensure the security function provided by the equipment is immediately restored. Physical security programs should be audited periodically to be certain requirements are being met.

Contingency Planning

Plant operators and facility managers must plan and provide for all reasonable contingencies, such as strikes, natural disasters, and onsite emergencies that require quick support and coordination from outside agencies and security personnel. Backup personnel should be identified to respond to these unusual situations. These plans should be reviewed periodically to ensure that contact information is current.

BIBLIOGRAPHY

All-Hazard Consequence Management Planning for the Water Sector. 2009. Critical Infrastructure Partnership Advisory Council Working Group, November 2009.

APHA, AWWA, and WEF (American Public Health Association, American Water Works Association, and Water Environment Federation). 2012. *Standard Methods for the Examination of Water and Wastewater*, 22nd ed. Washington, D.C.: American Public Health Association.

AWWA. 2002. M3—*Safety Practices for Water Utilities.* Denver, Colo.: American Water Works Association.

AWWA. 2001. M19—*Emergency Planning for Water Utilities.* Denver, Colo.: American Water Works Association.

AWWA and DHS. 2008. *Roadmap to Secure Control Systems in the Water Sector.* Water Sector Coordinating Council Cyber Security Working Group. March 2008.

AWWA. *Utility Security*, website resources, Denver, Colo.: American Water Works Association. http://www.awwa.org/legislation-regulation/issues/utility-security.aspx

AWWA. Water/Wastewater Agency Response Network (WARN). Denver, Colo.: American Water Works Association. http://www.awwa.org/resources-tools/water-knowledge/emergency-preparedness/water-wastewater-agency-response-network.aspx.

Code of Federal Regulations. OSHA Regulations. Occupational Safety and Health Standards, Part 1910. Washington, D.C.: OSHA.

Code of Federal Regulations. OSHA Regulations. Safety and Health Regulations for Construction, Part 1926. Washington, D.C.: OSHA.

Raimann, M. 2001. Cleveland Division of Water. Internal communication to author. September.

Sundheimer, M. 2001. Santa Fe Water Department. Internal communication to author. July.

The Chlorine Institute, 2008. Water and Wastewater Operators' Chlorine Handbook, Pamphlet 155, ed. 1. Washington, D.C.: The Chlorine Institute.

USEPA. 2010. *Water Sector–Specific Plan: An Annex to the National Infrastructure Protection Plan.* Washington D.C.: Department of Homeland Security.

Chapter 15

Record Keeping and Reporting

Records of plant operations and maintenance activities provide a basis for predictive decision-making and proof of compliance. Records also help operators to efficiently use chemicals (minimize waste) and equipment.

PROCESS RECORDS

In addition to the records that are required by the local regulatory agency, a water treatment plant should maintain records of activities as they relate to each process or discipline. Refer to AWWA's Principles and Practices of Water Supply Operations series for additional information on this topic.

The following sections describe the types of records operators should keep when performing their duties.

Coagulation and Flocculation

Maintain records of past performances for coagulants, including dosages, turbidities (sedimentation basin and filtration), particle counts, filterability indices, and zeta potentials as well as the temperatures, pH values, and color values at which the coagulants performed. For softening plants, record which softening chemicals are used (see chapter 9). Also record detention times and overflow rates in basins used for softening as well as residual quantities produced. Keep equipment records such as hours run for each mixer and flocculator, basin-cleaning activities, maintenance performed on motors and gears, and electrical use records.

Sedimentation

Calculate and record surface overflow rate, weir overflow rate, turbidity performance of each basin, and quantity of sludge removed from each basin (with a calculation for percent solids).

Maintain records of any operations and maintenance activities for each unit, including blowdown, cleaning, raking, and so on. Safety records for confined space entry may be necessary.

The Partnership for Safe Water suggests that the turbidity performance (and particle count, if available) in each sedimentation basin be recorded periodically (daily, every 4 hours, or hourly). Record the settled water turbidity as a combined flow, but realize that it cannot characterize the performance of an individual basin.

Filtration

For each filter, record the rate of flow, in million gallons per day or gallons per minute; head loss; length of run and unit filter run volume (UFRV); turbidity (in accordance with the Interim Enhanced Surface Water Treatment Rule [IESWTR]); particle counts; and dosage of any filter aid used (Table 15-1). At each backwash, record the amount of backwash water and surface wash water used, the length of the backwash, any observations during backwash, and time of filter-to-waste if used. Record the percentage of water treated that the backwash water represents (make note of amounts greater than 4 percent). All information can be included on one form.

On a quarterly basis, an operator should record bed expansion measurements, condition and depth of media, and an evaluation of the bed surface. On a yearly basis, record the results of solids retention analysis before and after backwash and the results of sieve analysis of the media. Record the addition of any media to the bed, including amounts and types. Keep a record of underdrain inspections and any work that was done on any appurtenances.

Record results of any testing required for filter-exceptions reporting under the IESWTR (this is regulated). Many states have required forms for plant operators to submit for filter performance. Consult the state regulatory agency for specific requirements.

Chlorination or Disinfection

Maintain records of the type of disinfectant used, including ordering information (phone number, address, and shipment amounts and container types), as well as information on costs and current dosage rates. Also results of bacteriological testing should be recorded for each process that uses a disinfectant, the water temperature and pH, and any unusual conditions that may occur. Keep records of safety training.

Precipitative Softening

Record the amount of softening chemicals used and the results of jar testing and lab analyses that justify these amounts. Record the results of analysis for alkalinity, pH, hardness, and related parameters for each shift, preferably every two hours for raw water and for all steps in the process. Keep a record of the chemical feeder setting and record the amount of water treated. Put everything on one form and maintain it daily.

Table 15-1 A sample filter record

Time	Rate of Flow, *mgd or gpm*	Head Loss, *ft*	Hours in Service	Total Volume Filtered (cumulative)	UFRV, *gal/ft²*	Turbidity, *ntu and particle counts*	Backwash Start	Length of Backwash, *minutes and amount used (gal)*	Surface Wash, *amounts and minutes*	Filter Aid Applied?, *Dosage?*	Filter to Waste? ntu at Startup
0000											
0200											
0400											
0600											
0800											
1000											
1200											
1400											
1600											
1800											
2000											
2200											

Record the amounts of sludge produced. If a solids contact unit is used, record the sludge blanket depth and results of settling tests. Keep records of the amounts and rates of solids returned to the process, if applicable.

Ion Exchange

Record the hardness, alkalinity, magnesium, calcium, and pH of the source and treated waters. Record the total amount of water treated and the amount of water treated by each softening unit, including bypass water. Also, keep a record of the amount of salt used for regeneration and the length of each cycle between regenerations. Record the amount of backwash water used and the amount of waste sent to sewer or other disposal places.

Aeration

Keep process records of raw water and finished water quality to determine if process is accomplishing its goals. Also, record the daily quantity of water treated, the details of safety and maintenance procedures, and changes in other treatment processes that may be affected by aeration. Document the process operating conditions including the air-to-water ratio to calculate performance and cost information.

Adsorption

Records kept regarding activated carbon differ depending on the type of carbon used.

- *Powdered activated carbon (PAC)*: Record the types of tastes and odors being experienced, the threshold odor number (TON), and the dates of taste-and-odor occurrences. Keep a record of the coagulant and PAC dosage predicted from jar tests as well as the actual dosage used.
- *Granular activated carbon (GAC)*: Record the number of filter hours; UFRV; any losses of GAC from the filter, measured monthly; dates of installation; and periodic TON from each filter. Also record when backwashing was performed and the amount of backwash water used as well as the raw-to-finished organic content removal.

Iron and Manganese Removal

If using ion exchange, the previously referenced records can be used. In addition, record raw and finished iron and manganese levels and results of distribution systems analyses performed. If permanganate is used, record amounts and dosages, including those predicted by jar testing.

If a sequestrant is used to control iron and manganese, record the amount and type used and the data from distribution system flushing efforts.

Fluoridation

Keep a daily record of the amount of fluoride in the finished water and the raw water, if necessary (a weekly raw water analysis may be sufficient if the regulator allows). Also, keep a daily record of the chemical dosages used and record the chemical feed rate each hour of each shift. Compute and record the daily fluoride chemical feed and compare it to the lab analysis as a double-check. Record safety classes and lectures that were provided to operators.

Corrosion Control

Maintain records of corrosion chemical dosages and any data available as to their performance, such as coupon testing results, Corrator® readings, temperatures, flushing activities, and lead and copper compliance results, i.e., 90th percentile levels and maximum levels. If phosphates are used, keep strict records of dosages and amounts of total and dissolved phosphate in the system. Flush mains regularly, especially dead-end mains; sample for phosphates at 5-, 10-, and 15-min intervals into the flush; and record results. Record the main flushing velocities, in feet per second. Track any customer complaints that may relate to use of corrosion chemicals.

Water Main Flushing

Main flushing records are best kept in a searchable database. Record main location and size; size of flushing pipe or hydrant; flow rate at flush; gallons used for flushing; velocity, in feet per second; chlorine residual; amount of phosphate used, if any; and results of biological testing, such as heterotrophic plate count quantity or quantity of coliform. Query the database by main size and show gallons used and time needed, along with any chemical and biological data, to generate monthly reports. In this way, a profile will emerge that shows the amount of water necessary to flush each size main, the time it takes, and the results obtained.

REPORTING

Reporting requirements are generally a function of the compliance reports that local and federal regulators require. In general, compliance testing results must be reported no later than 10 days after the end of the month in which they were accumulated. It is good practice for water treatment plant staff to provide reports to upper management. Even if the administration does not require these reports, they should be generated and kept on file for future reference.

Like a diary, a written report is an excellent reminder of events that took place and the methods used to handle these events. Internal memos of safety issues, personnel and disciplinary actions, purchases, and maintenance events can actually help protect staff from allegations of nonfeasance.

Reports should be written as soon as possible after an event takes place so that details can be easily recalled. Do not write reports or internal memos in anger; take time to cool down before recalling an event.

Most compliance reports must conform to a specific format. This enables the regulator to assemble large amounts of data from many utilities in an orderly fashion. Internal or outsourced reports, other than compliance reports to regulators, should include the following:

- Times and dates of the audit and any description of incidents or accomplishments
- Financial or inventory considerations
- Personnel involved
- Conclusions drawn, with supporting data and references
- List of report recipients, with acknowledgment of receipt, if appropriate
- An executive summary if the report is lengthy and involved

The report should be filed in a logical order, e.g., file reports under general categories such as research, safety, finance, and so on, so that the report can be easily retrieved in the future. As time permits, reread all reports to gauge progress in reaching goals and to learn from past mistakes.

Reporting Treatment Incidents

There are specific requirements for the time and method of reporting a regulatory violation. These are described in the drinking water regulations for the applicable regulatory agency. However, there are times when it is not clear if there has been a violation or when there is a water quality incident that may affect consumers. The water plant personnel should have a predetermined plan of action to deal with these situations.

The notification and action plan for incident reporting should include plant management, utility management, other utility officials, and regulatory agencies. The utility should have discussions with the regulatory agency representatives prior to any incident to establish an understanding of how they will handle this type of communication. The agency may have technical resources to assist the plant personnel with a response. Also, the agency may be helpful when dealing with mass communication media regarding the incident.

PLANT PERFORMANCE REPORTS

Periodic (monthly, quarterly, and annual) plant performance reports should be prepared and these should be examined to detect trends and identify opportunities for improvement. The operating reports previously described are the basis for these performance reports. Performance reports summarize the operating results and usually include various charts and calculated performance measures. These summary reports do not need to be lengthy and, indeed, can be added to the operating reports if desired. The purpose of the performance reports is to examine the plant results over longer time periods to see if there are trends that need investigation. All plant operating personnel should review these reports to provide feedback and to participate in efforts to meet plant operating goals.

BIBLIOGRAPHY

AWWA. 2010. Principles and Practices of Water Supply Operations—*Water Treatment*, 4th ed. Denver, Colo.: American Water Works Association.

AWWA and AwwwaRF. 1997. *Partnership for Safe Water Self-Assessment Guide for Surface Water Treatment Optimization*. Denver, Colo.: American Water Works Association and American Water Works Association Water Research Foundation.

USEPA. Reporting Requirements for Regulations. Washington D.C.: USEPA http://www.epa.gov/lawsregs/topics/water.html

Appendix A

Sample Material Safety Data for Chlorine

PRODUCT NAME: CHLORINE

1. Chemical Product and Company Identification

BOC Gases,
Division of
The BOC Group, Inc.
575 Mountain Avenue
Murray Hill, NJ 07974

TELEPHONE NUMBER: (908) 464-8100
24-HOUR EMERGENCY TELEPHONE NUMBER:
CHEMTREC (800) 424-9300

BOC Gases
Division of
BOC Canada Limited
5975 Falbourne Street, Unit 2
Mississauga, Ontario L5R 3W6
TELEPHONE NUMBER: (905) 501-1700

24-HOUR EMERGENCY TELEPHONE NUMBER:
(905) 501-0802
EMERGENCY RESPONSE PLAN NO: 2-0101

PRODUCT NAME: CHLORINE
CHEMICAL NAME: Chlorine
COMMON NAMES/SYNONYMS: Bertholite, Molecular Chlorine
TDG (Canada) CLASSIFICATION: 2.3 (5.1)
WHMIS CLASSIFICATION: A, D1A, D2B, E, C

PREPARED BY: Loss Control (908)464-8100/(905)501-1700
PREPARATION DATE: 6/1/95
REVIEW DATES: 4/16/02

2. Composition, Information on Ingredients

EXPOSURE LIMITS[1]:

INGREDIENT	% VOLUME	PEL-OSHA[2]	TLV-ACGIH[3]	LD_{50} or LC_{50} Route/Species
Chlorine FORMULA: Cl_2 CAS: 7782-50-5 RTECS #: FO2100000	100.0	1 ppm Ceiling	0.5 ppm TWA 1 ppm STEL	LC_{50}: 293 ppm inhalation/rat (1H)

[1] Refer to individual state or provincial regulations, as applicable, for limits which may be more stringent than those listed here.
[2] As stated in 29 CFR 1910, Subpart Z (revised July 1, 1993)
[3] As stated in the ACGIH 2002 Threshold Limit Values for Chemical Substances and Physical Agents.

OSHA Regulatory Status: This material is classified as hazardous under OSHA regulations.
IDLH: 10 ppm

3. Hazards Identification

EMERGENCY OVERVIEW

Greenish yellow gas with bleach-like choking odor. Corrosive and poison gas. Contact may cause severe irritation or corrosive burns to the eyes, skin and mucous membranes. Inhalation may result in chemical pneumonitis, retention of body fluid in the lungs (pulmonary edema), and respiratory collapse. Nonflammable. Oxidizer. May react violently with reducing agents. Can accelerate combustion and increase risk of fire and explosion in flammable and combustible materials. Contents under pressure. Use and store below 125°F.

Source: BOC Gases, a Division of The BOC Group Inc.

ROUTE OF ENTRY:

Skin Contact	Skin Absorption	Eye Contact	Inhalation	Ingestion
Yes	No	Yes	Yes	No

HEALTH EFFECTS:

Exposure Limits	Irritant	Sensitization
Yes	Yes	No
Teratogen	Reproductive Hazard	Mutagen
No	No	No
Synergistic Effects Other agents that irritate the respiratory system		

Carcinogenicity: -- NTP: No IARC: No OSHA: No

EYE EFFECTS:
Corrosive and irritating to the eyes. Contact with the liquid or vapor causes painful burns and ulcerations. Burns to the eyes result in lesions and possible loss of vision.

SKIN EFFECTS:
Corrosive and irritating to the skin and all living tissue. It hydrolyzes very rapidly yielding hydrochloric acid. Skin burns and mucosal irritation are like that from exposure to volatile inorganic acids. Chlorine burns result in severe pain, redness, possible swelling and early necrosis.

INGESTION EFFECTS:
Ingestion is unlikely.

INHALATION EFFECTS:
Corrosive and irritating to the upper and lower respiratory tract and all mucosal tissue. Symptoms include lacrimation, cough, labored breathing, and excessive salivary and sputum formation. Excessive irritation of the lungs causes acute pneumonitis, pulmonary edema, and respiratory collapse which could be fatal. Residual pulmonary malfunction may also occur. Chemical pneumonitis and pulmonary edema may result from exposure to the lower respiratory tract and deep lung.

MEDICAL CONDITIONS AGGRAVATED BY EXPOSURE: May aggravate pre-existing eye, skin, and respiratory conditions.

POTENTIAL ENVIRONMENTAL EFFECTS: Toxic to fish and wildlife. Chlorine is designated as a marine pollutant by DOT. The LC_{50} in the fathead minnow has been cited as 0.1 mg/l/96 H and an LC_{50} of 0.097 mg/L/30 min has been cited for the *Daphnia magna*.

4. First Aid Measures

EYES:
PERSONS WITH POTENTIAL EXPOSURE SHOULD NOT WEAR CONTACT LENSES. Flush contaminated eye(s) with copious quantities of water. Part eyelids to assure complete flushing. Continue for a minimum of 30 minutes. Seek immediate medical attention.

SKIN:
Flush affected area with copious quantities of water while removing contaminated clothing. Seek immediate medical attention.

INGESTION:
None required.

INHALATION:
PROMPT MEDICAL ATTENTION IS MANDATORY IN ALL CASES OF OVEREXPOSURE. RESCUE PERSONNEL SHOULD BE EQUIPPED WITH SELF-CONTAINED BREATHING APPARATUS. Conscious persons should be assisted to an uncontaminated area and inhale fresh air. If breathing is difficult, administer oxygen. Unconscious persons should be moved to an uncontaminated area and given artificial resuscitation and supplemental oxygen. Assure that mucus or vomited material does not obstruct the airway by use of positional drainage. Delayed pulmonary edema may occur. Keep the patient under medical observation for at least 24 hours.

5. Fire Fighting Measures

<table>
<tr><td colspan="4">Conditions of Flammability: Not flammable</td></tr>
<tr><td>Flash point:
None</td><td colspan="2">Method:
Not Applicable</td><td>Autoignition
Temperature: None</td></tr>
<tr><td colspan="2">LEL(%): None</td><td colspan="2">UEL(%): None</td></tr>
<tr><td colspan="4">Hazardous combustion products: None</td></tr>
<tr><td colspan="4">Sensitivity to mechanical shock: None</td></tr>
<tr><td colspan="4">Sensitivity to static discharge: None</td></tr>
</table>

FIRE AND EXPLOSION HAZARDS:
Strong oxidizer. Most combustible materials burn in chlorine as they do in oxygen producing irritating and poisonous gases. Flame impingment upon steel chlorine container will result in iron/chlorine fire causing rupture of the container. Cylinder may vent rapidly or rupture violently from pressure when involved in a fire situation.

EXTINGUISHING MEDIA:
Use media suitable for surrounding materials. If it can be done without risk, stop the flow of chlorine which is accelerating the fire.

FIRE FIGHTING INSTRUCTIONS:
Firefighters should wear respiratory protection (SCBA) and full turnout or Bunker gear with additional chemical protective clothing to prevent exposure to chlorine. Use water spray to keep fire exposed containers cool. Continue to cool fire exposed cylinders until well after flames are extinguished. Control runoff and isolate discharged material for proper disposal.

6. Accidental Release Measures

Evacuate all personnel from affected area. Deny entry to unauthorized and unprotected individuals. Extinguish all ignition sources. No smoking, sparks, flames, or flares in hazard area. Appropriate protective equipment is essential to prevent exposure (See Section 8). Stop the flow of gas or remove cylinder to outdoor location if this can be done without risk. Ventilate enclosed areas. A leak near incompatible, flammable or combustible materials may create a fire or explosion hazard. Consult a HAZMAT specialist and the appropriate emergency telephone number in Section 1 or your closest BOC location. If leak is in user's equipment, be certain to purge piping with inert gas prior to attempting repairs.

7. Handling and Storage

Electrical classification: Nonhazardous.

Most metals corrode rapidly with wet chlorine. Systems must be kept dry. Lead, gold, tantalum and Hastelloy are most resistant to wet chlorine.

Do not inhale. Prevent contact with skin and eyes. Use only in well-ventilated areas. Valve protection caps must remain in place unless container is secured with valve outlet piped to use point. Do not drag, slide or roll cylinders. Use a suitable hand truck for cylinder movement. Use a pressure reducing regulator when connecting cylinder to lower pressure piping or systems. Do not heat cylinder by any means to increase rate of product from the cylinder. Use a check valve or trap in the discharge line to prevent hazardous back flow into cylinder. Do not insert any object (i.e.: screwdriver) into valve cap openings as this can damage the valve causing leakage.

Protect cylinders from physical damage. Store in cool, dry, well-ventilated areas of non-combustible construction away from heavily trafficked areas and emergency exits. Do not allow the temperature where cylinders are stored to exceed 125°F (52°C). Cylinders should be stored upright and firmly secured to prevent falling or being knocked over. Full & empty cylinders should be segregated. Use a "first in-first out" inventory system to prevent full cylinders from being stored for excessive periods of time. Separate from combustibles, organic, and easily oxidizable materials. Isolate from acetylene, ammonia, hydrogen, hydrocarbons, ether, turpentine, finely divided metals, and other incompatible materials. Outside or detached storage is preferred.

Never carry a compressed gas cylinder or a container of a gas in cryogenic liquid form in an enclosed space such as a car trunk, van or station wagon. A leak can result in a fire, explosion, asphyxiation or a toxic exposure.

For additional storage recommendations, consult Compressed Gas Association's Pamphlet P-1.

8. Exposure Controls, Personal Protection

ENGINEERING CONTROLS:
Hood with forced ventilation may be used for small quantities. Use local exhaust ventilation in combination with enclosed processes as needed to prevent accumulation above the exposure limit.

EYE/FACE PROTECTION:
Gas-tight safety goggles and full faceshield or full-face respirator.

SKIN PROTECTION:
Protective gloves or fully encapsulated vapor protective clothing. (Butyl rubber, neoprene, and Teflon ® provide adequate protection for exposures to chlorine greater than 8 hours.)

RESPIRATORY PROTECTION:
For emergency release use a positive pressure NIOSH approved air-supplying respirator systems (SCBA or airline/escape bottle) using a full-face mask and at a minimum Grade D air.

For normal conditions below fifty times the exposure limit but where engineering can not control exposures below the applicable limits, than appropriately selected air-purifying respirators with full-face mask can be used.

OTHER/GENERAL PROTECTION:
Safety shoes, safety shower, eyewash "fountain"

9. Physical and Chemical Properties

PARAMETER	VALUE	UNITS
Physical state (gas, liquid, solid)	: Gas	
Vapor pressure at 70 °F	: 100.2	psia
Vapor density at STP (Air = 1)	: 2.47	
Evaporation point	: Not Available	
Boiling point	: -29.3	°F
	: -34.1	°C
Freezing point	: -149.8	°F
	: -101	°C
pH	: Not Available	
Specific gravity	: Not Available	
Oil/water partition coefficient	: Not Available	
Solubility (H_20)	: Very Soluble	
Odor threshold	: Not Available	
Odor and appearance	: Greenish-yellow gas with sharp suffocating odor. Liquid is amber colored.	

10. Stability and Reactivity

STABILITY: Stable

INCOMPATIBLE MATERIALS: Strong oxidizer. Will react with organic and other oxidizable materials. Reacts explosively or forms explosive compounds with many common substances including acetylene, ether, turpentine, ammonia, fuel gas, hydrogen and finely divided metals. Reacts with water to form corrosive acidic solution.

HAZARDOUS DECOMPOSITION PRODUCTS: Hydrochloric acid on contact with water.

HAZARDOUS POLYMERIZATION: Will not occur.

11. Toxicological Information

INHALATION:
Inhalation of chlorine concentrations as low as 1 ppm may cause nose, throat and conjunctiva irritation. Irritation becomes more pronounced at concentrations of 1.3 ppm and above with coughing and labored breathing. Death may occur after a few breaths at 1000 ppm. Delayed effects following high exposure may include bronchitis, edema, and pneumonia.

SKIN AND EYE:
Extremely irritating to the skin, eyes, and mucous membranes. Can cause corrosive burns. May cause corrosion of the teeth. Prolonged exposure to low concentrations may cause chloracne.

CHRONIC:
Repeated contact with low concentrations may cause dermatitis.

OTHER:
Equivocal evidence of carcinogenicity for chlorine was noted in an IARC review and a 2-year drinking water study in F344/N rats and B6C3F1 mice by the NTP. Literature references suggest the possibility of mutagenic and teratogenic effects from hypochlorites (a hydrolysis product of chlorine).

12. Ecological Information

Product does not contain Class I or Class II ozone depleting substances. Chlorine is highly toxic to all forms of aquatic life (See Section 3). There is no potential for bioaccumulation or bioconcentration. Chlorine is designated as a hazardous substance under section 311(b)(2)(A) of the Federal Water Pollution Control Act and further regulated by the Clean Water Act Amendments of 1977 and 1978. Listed as a hazardous air pollutant (HAP) and a marine pollutant. Chlorine is listed as an extremely hazardous substance (EHS) subject to state and local reporting under Section 304 of SARA Title III (EPCRA) with a Threshold Planning Quantity (TPQ) of 100 pounds. The CERCLA reportable quantity (RQ) for chlorine is 10 pounds.

13. Disposal Considerations

Do not attempt to dispose of residual waste or unused quantities. Return in the shipping container PROPERLY LABELED, WITH ANY VALVE OUTLET PLUGS OR CAPS SECURED AND VALVE PROTECTION CAP IN PLACE to BOC Gases or authorized distributor for proper disposal.

14. Transport Information

PARAMETER	United States DOT	Canada TDG
PROPER SHIPPING NAME:	Chlorine	Chlorine
HAZARD CLASS:	2.3 (8)	2.3 (8)
IDENTIFICATION NUMBER:	UN 1017	UN 1017
SHIPPING LABEL:	POISON GAS, CORROSIVE	TOXIC GAS, CORROSIVE

* Effective August 15, 2002

Additional Marking Requirement: "Inhalation Hazard"

If net weight of product ≥ 10 pounds, the container must be also marked with the letters "RQ".
"Marine Pollutant" – For vessel transportation the Marine Pollutant Mark shall be placed in association with the hazard warning labels, or in the absence of any labels, in association with the marked proper shipping name.

Additional Shipping Paper Description Requirement: "Poison-Inhalation Hazard, Zone B"

If net weight of product ≥ 10 pounds, the shipping papers must be also marked with the letters "RQ".
The words "Marine Pollutant" shall be entered in association with the basic description for a material which is a marine pollutant.

15. Regulatory Information

SARA TITLE III NOTIFICATIONS AND INFORMATION
SARA TITLE III - HAZARD CLASSES:
Acute Health Hazard
Chronic Health Hazard
Fire Hazard
Sudden Release of Pressure Hazard
Reactivity Hazard

SARA TITLE III - SECTION 313 SUPPLIER NOTIFICATION:
This product contains the following toxic chemicals subject to the reporting requirements of section 313 of the Emergency Planning and Community Right-To-Know Act (EPCRA) of 1986 and of 40 CFR 372:

CAS NUMBER	INGREDIENT NAME	PERCENT BY VOLUME
7782-50-5	CHLORINE	100.0

This information must be included on all MSDSs that are copied and distributed for this material.

U.S. TSCA/Canadian DSL: All ingredients are listed on the U.S. Toxic Substances Control Act (TSCA) inventory or exempt from listing and on the Canadian Domestic Substance List (DSL).

California Proposition 65: This product does not contain ingredient(s) known to the State of California to cause cancer or reproductive toxicity.

16. Other Information

NFPA HAZARD CODES		HMIS HAZARD CODES		RATINGS SYSTEM
Health:	4	Health:	3	0 = No Hazard
Flammability:	0	Flammability:	0	1 = Slight Hazard
Instability:	0	Physical Hazard:	2	2 = Moderate Hazard
OXIDIZER				3 = Serious Hazard
				4 = Severe Hazard

ACGIH	American Conference of Governmental Industrial Hygienists
DOT	Department of Transportation
IARC	International Agency for Research on Cancer
NTP	National Toxicology Program
OSHA	Occupational Safety and Health Administration
PEL	Permissible Exposure Limit
SARA	Superfund Amendments and Reauthorization Act
STEL	Short Term Exposure Limit
TDG	Transportation of Dangerous Goods
TLV	Threshold Limit Value
WHMIS	Workplace Hazardous Materials Information System

Compressed gas cylinders shall not be refilled without the express written permission of the owner. Shipment of a compressed gas cylinder which has not been filled by the owner or with his/her (written) consent is a violation of transportation regulations.

DISCLAIMER OF EXPRESSED AND IMPLIED WARRANTIES:
Although reasonable care has been taken in the preparation of this document, we extend no warranties and make no representations as to the accuracy or completeness of the information contained herein, and assume no responsibility regarding the suitability of this information for the user's intended purposes or for the consequences of its use. Each individual should make a determination as to the suitability of the information for their particular purpose(s).

Appendix B

Math and Calculation Methods

PRACTICAL UNIT CONVERSIONS FOR WATER

Cubic feet (ft^3) to gallons (gal)

1 ft^3 of water is equivalent to 7.48 gal at normal temperatures.

Example: If a residential water meter registers in units of 100 ft^3 and 25 units of water have passed through the meter, how many gallons is this? 1 ft^3 is 7.48 gal, so 25 × 100 × 7.48 gal/ft^3 = 18,700 gal.

Gallons (gal) of water to pounds (lb)

1 gal of water weighs approximately 8.34 lb

Example: How many pounds of water are contained in a 50-gal day tank? 50 × 8.34 = 417 lb

Gallons (gal) to liters (L)

1 gal is equivalent to 3.785 L, or 3,785 mL.

Example: An operator transfers 36 gal of polymer to a day tank. How many liters is this? 1 gal is 3.785 L, so 36 × 3.785 = 136.26 L. This is 136,260 mL.

Grams (g) to pounds (lb)

1 lb is equivalent to 453.6 g.

Example: An operator feeds 13.5 lb of chemical each hour. How many grams is this? 453.6 × 13.5 = 6,123.6 g

Million gallons per day (mgd) to cubic feet per second (ft^3/sec)

1 mgd is equivalent to 1.55 ft^3/sec.

Example: What is the average flow rate, in cubic feet per second, in a plant that produces 20 mgd? 1 mgd = 1.55 ft^3/sec, so 20 mgd × 1.55 = 31 ft^3/sec

Million gallons per day (mgd) to gallons per minute (gpm)

1 mgd is equivalent to 694 gpm.

Example: If a water treatment plant is operating at 6.75 mgd, how many gallons per minute is this? 1 mgd = 694 gpm, so 6.75 × 694 = 4,684.5 gpm

Cubic feet per second (ft^3/sec) to gallons per minute (gpm)

1 ft^3/sec = 448.8 gpm.

Example: How many gallons per minute are represented in a flow of 3 ft^3/sec? 3 ft^3/sec × 60 sec/min = 180 ft^3/min; 180 ft^3/min × 7.48 gal = 1,346.4 gpm

Percent (%) concentration to milligrams per liter (mg/L)

1% is equivalent to 10,000 mg/L.

Example: An operator makes a 0.75% polymer solution in a day tank. How many milligrams per liter is this? 1% = 10,000 mg/L, so 0.75% × 10,000 = 7,500 mg/L

Feet (ft) of head to pounds per square inch (psi) pressure

1 psi is equivalent to a column of water that stands vertically at 2.31 ft.

Example: How many feet of head are developed at the base of a water tank that is 135 ft high? 2.31 ft of head = 1 psi, so 135 ft/2.31 = 58.4 psi.

Pounds per square inch (psi) pressure to feet (ft) of head

1 ft of head is equivalent to 0.433 psi.

Example: How many feet of head are developed in a pumping system that is registering 68 psi? 1 ft of head = 0.433 psi, so 68 psi divided by 0.433 = 157 ft.

Grains per gallon (gpg) to milligrams per liter (mg/L)

1 gpg = 17.1 mg/L

Example: What is a water hardness of 11 gpg in mg/L? 11gpg x 17.1 mg/L for each gpg = 188.1 mg/L hardness.

Time conversions

1 day is equivalent to 24 hours, which is equivalent to 1,440 minutes. 1 minute is equivalent to 60 seconds. There are 86,400 seconds in 1 day.

PRACTICAL WATER TREATMENT PLANT EXAMPLE PROBLEMS

Volume

What is the volume, in gallons, of a rectangular clearwell that is 38 ft × 75 ft if the water depth is 14 ft? 38 ft × 75 ft × 14 ft × 7.48 gal/ft^3 = 298,452 gal

How many cubic feet of water are contained in 1,000 ft of 8-in. water main? Volume = $\pi R^2 H$ = 3.14 × 0.33 ft × 0.33 ft × 1,000 = 341.95 ft^3

What is the volume, in gallons, of a 60-ft-diameter tank when the water depth is 30 ft? Volume = 0.785 × (D^2) × depth × 7.48 gal/ft^3 = 0.785 × (60 ft)(60 ft) × 30 ft × 7.48 = 634,154 gal

Solution strength

An operator has a 10% stock solution of polymer and wishes to make 30 gal of a 1.5% working solution from the stock. How many gallons of the stock are needed? (30 gal)(15,000 mg/L) = (x gal)(100,000 mg/L), therefore x = 4.5 gal.

If 500 lb of a 4.5% solution is mixed with 800 lb of a 12% solution, what is the resultant solution strength? (500 lb × .045) + (800 lb × .12)/(500 + 800) = 9.1%, or 91,000 mg/L

Feed rate

The operator makes the 1.5% working solution of polymer in the previous example. If the operator wants to treat a flow of 1,200 gpm with 0.5 mg/L of the polymer, how many milliliters per minute are needed? (1,200 gpm)(0.5 mg/L) = (x gpm)(15,000 mg/L), so x = 0.04 gpm feed needed. And, 0.04 gpm × 3.785 L/gal = 0.1514 L/min, or 151 mL/min.

An operator feeds activated carbon at a rate of 23 lb/hr to treat a raw flow of 32 mgd. How many milligrams per liter is fed? 23 lb/hr × 24 hr = 552 lb/d. 552 lb/d /32 mgd = 17.25 lb/mil gal. 17.25 lb/mil gal/8.34 = 2 mg/L.

The gas chlorinators at a water plant go out of service; the operator decides to feed a hypochlorite solution that is 15% as available chlorine. If the operator needs to treat a flow of 1.6 mgd with 6 mg/L chlorine, how many gallons per hour hypochlorite solution are needed? 1,600,000 gpd/24 hr = 67,000 gph. (67,000 gph × 6 mg/L) / 150,000 mg/L = 2.68 gph.

At a water treatment plant operating at 4 mgd, the dry alum feeder is set at an indicated feed rate of 17.3 mg/L. The operator runs a calibration check by catching 0.34474 kg of dry alum in 2 min. Is the dosimeter accurate? 0.34474 × 1,000 = 344.74 g. (344.74 g/2 min) × (1 lb/453.6 g) = 0.38 lb/min actually fed. The dosimeter is set at 17.3 mg/L, which is (17.3 × 8.34 × 4 mgd)/1,440 min = 0.40 lb/min. The dosimeter is not accurate.

Detention time

Detention time = V/Q, where V is the volume in cubic feet and Q = the flow in cubic feet per second or per minute or per hour.

If a 4.52-mgd pump is used to fill a rectangular basin that is 75 ft × 36 ft, how long will it take to bring the water level from empty to 8 ft? V = 75 ft × 36 ft × 8 ft = 21,600 ft^3. Q = 4.52 mgd × 1.55 = 7 ft^3/sec. Therefore, DT = 21,600 ft^3/ 7 ft^3/sec = 3085.7 sec, or 51.4 min.

What is the carbon contact time if powdered activated carbon is fed into a 36 inch diameter raw water force main 23,000 ft in length and the flow through the force main is 8 MGD? 8,000,000 gal/day x 1 day/ 24 hr x 1 ft^3 / 7.48 gal = 44,563 cubic feet per hour. DT = V/flow = (3.14 x 1.5 ft x 1.5 ft x 23,000 ft) / 44,563 ft^3 / hr = 3.64 hours

A wash water recycle basin that measures 25 ft × 80 ft × 15 ft is emptied by a recycle pump operating at 560 gpm. Filters are being backwashed at the same time, and the flow from the backwash enters the recycle basin. Each filter backwash uses 54,600 gal. How many filters can be backwashed over a 4-hour period, assuming the basin was empty from the start? Basin volume is 25 ft × 80 ft × 15 ft × 7.48 = 224,000 gal. Basin is emptied at 560 gpm for 4 hours, which is 560 gpm × 60 min × 4 hr = 134,000 gal. Therefore, effective volume of the recycle basin is 134,000 gal + 224,000 gal = 358,000 gal. 358,000 gal/54,600 gal/wash = 6.57 washes. Six filters can be washed, but not seven.

Well drawdown

The level of the aquifer of a well before pumping is 130 ft. When the pump is operated, the aquifer level drops to 173 ft. What is the drawdown? 173 ft – 130 ft = 43 ft

Well yield and specific yield of wells

A total of 950 gal were pumped from a well in a period of 5 minutes. What is the well yield in gpm? 950 gal/5 min = 190 gpm

Using the well yield in the previous example, calculate the specific yield if the drawdown is 25 ft. 190 gpm/25 ft = 7.6 gpm/ft

Disinfection of system components

An operator wishes to disinfect a new well casing that has a 12-in. diameter. The desired chlorine dosage is 50 mg/L. The casing is 150 ft long, and the water level in the casing is 55 ft from the top of the well. How many pounds of chlorine are needed? The water column is 150 ft – 55 ft = 95 ft. Therefore, the volume is 0.785 × 1 ft × 1 ft × 95 ft × 7.48 = 558 gal. Dosage = 50 mg/L × (558/1,000,000) × 8.34 = 0.23 lb chlorine.

Determine the amount of 65% calcium hypochlorite needed to disinfect the water tower and riser if a 50 mg/L dose is desired.

Head

The static water level of a well pump is 90 ft. The well drawdown is 29 ft. If the gauge reading at the pump discharge head is 3.9 psi, what is the total pumping head? (90 ft + 29 ft) + (3.9 psi × 2.31) = 128 ft

Using the information depicted in Figure B-1, determine the pressure at the main. Mercury is 13.6 times as heavy as water. The pressure will be the difference between the force created by the mercury and the force created by the water, so (3 × 13.6 ft) – (3 × 1 ft) = 37.8 ft of head. 37.8 ft × 0.433 = 16.4 psi

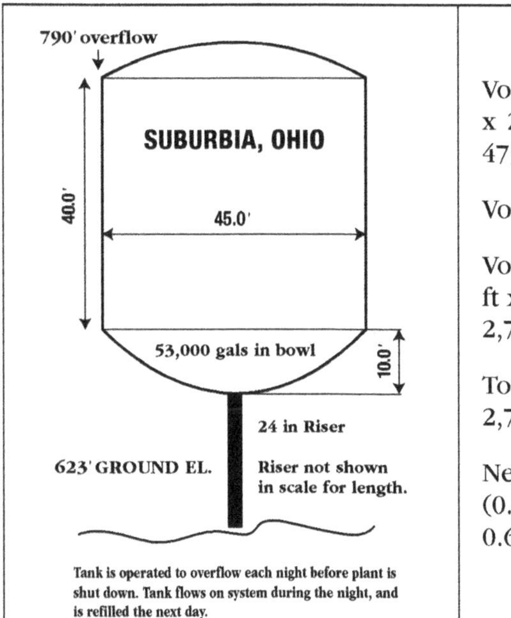

Volume of tank = 3.14 x 22.5 ft x 22.5 ft x 40 ft x 7.48 gal/ft^3 = 475,615 gal

Volume of bowl = 53,000 gal

Volume of riser = 3.14 x 1 ft x 1 ft x (740 ft-623 ft) x 7.48 gal/ft^3 = 2,748 gal

Total Volume = 475,615 + 53,000 + 2,748 = 531,363 gal, or 0.531 MG

Needed calcium hypochlorite = (0.521 MG x 50 mg/L x 8.34) / 0.65 = 340 lb

Figure B-1 Disinfection of water tower

Density and specific gravity

Density is usually expressed as weight per unit volume, such as pounds per gallon. Specific gravity is a ratio of the density of a liquid to the density of water. The density of water at room temperature is approximately 8.34 lb/gal. The density of water can vary with temperature; in fact, water is most dense at 39.4°F (4°C). If 1 gal of oil weighs 7.51 lb, what is the specific gravity? (7.51 lb/gal)/(8.34 lb/gal) = 0.90.

The specific gravity of a batch of liquid alum is 1.295. What is the density? 1.295 × 8.34 lb/gal = 10.8 lb/gal. Each gallon of alum fed, in this case, would equate to 10.8 lb of liquid alum.

The specific gravity of mercury is 13.6. What does 1 gal of mercury weigh? 13.6 × 8.34 = 113.4 lb

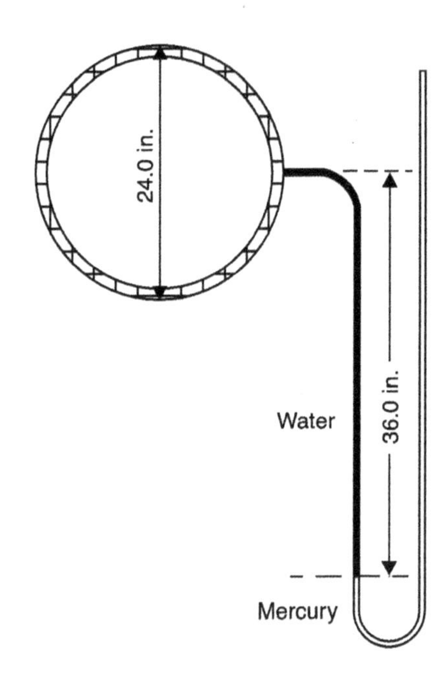

Figure B-2 **Mercury manometer setup on 24-in. main**

Flow rate and velocity

The flow rate calculation for water moving through a pipe or channel is $Q = A \times V$, where Q is flow rate in cubic feet per second, A is the square foot area of the pipe or channel, and V is the velocity in feet per second.

Newly installed water mains are normally flushed at 2.5 fps. How many cubic feet per second are needed to flush a 12-in.-diameter main? First, convert the 12-in. diameter to a radius in feet. 12 in./2 = 6 in. radius, and 6 in./12 = 0.5 ft. $Q = 3.14 \times 0.5$ ft $\times$ 0.5 ft $\times$ 2.5 fps = 1.96 ft^3/sec.

A water plant is operating at 3.6 mgd. After the flow travels through the rapid mix, it splits into two equally sized channels on its way to the flocculators. What is the velocity through one channel if it is 1.8 ft wide and the water depth is 18 in.? 3.6 mgd $\times$ 1.55 = 5.58 ft^3/sec. 5.58 ft^3/sec/2 = 2.79 ft^3/sec through each channel. V = Q/A, or V = 2.79 ft^3/sec/(1.8 ft $\times$ 1.5 ft) = 1 ft/sec.

What is the velocity in meters per second in a pipe with a diameter of 42 cm and a flow of 7,500 m^3/d? 42 cm diameter = 0.42 m diameter, and 0.42/2 = 0.21 m radius. V = Q/A = 7,500 m^3/d/(3.14 $\times$ 0.21 m $\times$ 0.21 m) = 54,161.8 m/d or 0.63 m/sec.

The velocity in an 8 in. water main is 2 fps. How many gpm are flowing through it? Diameter of main is 8 in., so radius in feet is (8 in. /1 ft/ 12 in.)/2 = 0.34 ft^2 area. Velocity is 2 ft/sec x 60 sec/min = 120 ft / min. Q = A x V = 0.34 ft^2 x 120 ft/min x 7.45 gal/ft^3 = 305 gpm.

Weir overflow and surface overflow rates

Weir overflow rates are expressed as gallons per day per foot, or gallons per minute per foot, while surface overflow rates are expressed as gallons per day per square foot or gallons per minute per square foot. Weir overflow rates use the total linear distance of the weirs at the end of the basins for a calculation, and surface overflow rates use the square foot area of the surface of the basin. Both formulas work for rectangular and circular basins, and neither formula depends on basin depth.

Calculate the weir loading in a plant that is operating at 8 mgd and has four sedimentation basins, each with a 200-ft weir. 8,000,000/(4×200 ft) = 10,000 gpd/ft

Calculate the surface overflow rate for the same plant if each basin is 38 ft $\times$ 106 ft. 8,000,000/(4 $\times$ 38 ft $\times$ 106 ft) = approximately 500 gpd/ft^2.

Filter loading and backwash rate

A water treatment plant operates four filters, each are 32 ft × 26 ft. What is the average filtration rate if the plant flow is 8 mgd? 8,000,000 gpd/1,440 = 5,555 gpm. (4)(32 ft × 26 ft) = 3,328 ft^2. 5,555 gpm/3,328 ft^2 = 1.67 gpm/ft^2.

An operator needs to calculate the average wash time for a filter that measures 14 ft × 26 ft. If the filter was washed 46 times, and 2,511,600 gal were used during that span of time at an average backwash rate of 15 gpm/ft^2, what was the average length of backwash? Area of filter is 14 ft × 26 ft = 364 ft^2. 2,511,600 gal/46 washes = 54,600 gal/wash. 15 gpm/ft^2 × 364 ft^2 = 5,460 gpm during the wash. (54,600 gal/wash)/(5,460 gpm) = 10 min/wash, on average.

Horsepower and Pumping

The power required to drive a pump is called water horsepower (WHP). Because pumps are inefficient, the power requirements for the pump are calculated as brake horsepower (BHP). The motors that drive pumps are also inefficient, and so that power requirement calculation is called motor horsepower (MHP) (see Figures B-3, B-4, and B-5).

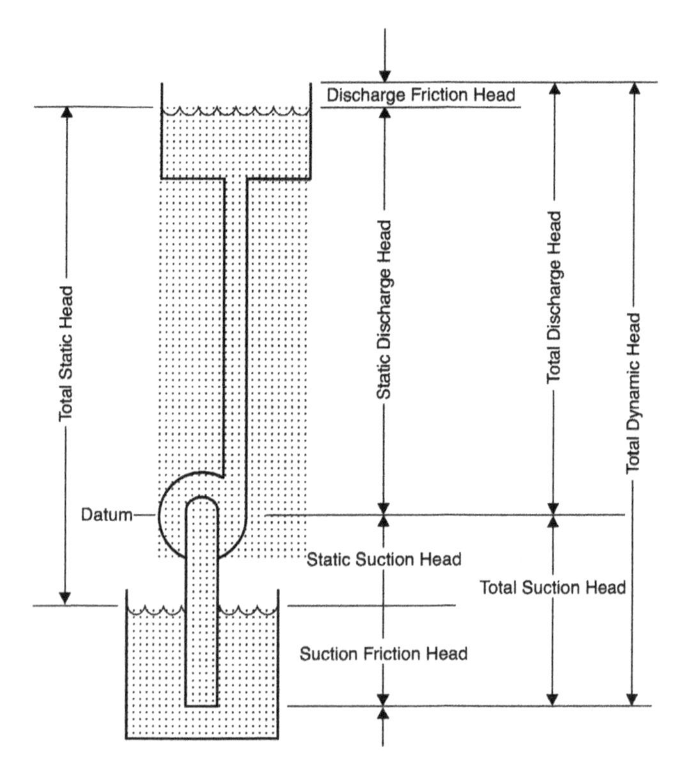

Figure B-3 **Schematic illustrating common pumping terminology**

Source: Philadelphia Water Department.

Figure B-4 Pump impeller maintenance. Impellers and shafts ready for cleaning.

Source: Cleveland, Ohio, Division of Water.

Figure B-5 Operator checking pump control. This is a 38-mgd high service pump operating at 2,250 hp.

What is the WHP requirement to pump 375 gpm against a head of 55 psi?

Formula for WHP = (gpm × head in feet)/3,960. Therefore, (375 × (55 × 2.31))/3,960 = 12 hp.

If the pump in the previous problem is 87% efficient, what is the BHP requirement? BHP = WHP/efficiency = 12/0.87 = 13.8 BHP.

If the motor that drives the pump in the previous problems is 92% efficient, what is the MHP requirement? MHP = BHP/efficiency = 13.8/0.92 = 15 MHP.

The previous examples show that it always takes more energy to move water through systems because inefficiencies result from friction and heat loss.

What is the WHP required to fill an elevated storage tank with a pump operating at 5,600 gpm if the static head is 85 psi and the suction head is 9 ft? 85 psi × 2.31 = 196.35 ft. 196.35 ft – 9 ft = 187.35 ft. Therefore, (5,600 gpm × 187.35) / 3,960 = 265 hp.

C × T

Using Figure B-6 and Table B-1, calculate the reportable C × T value in milligram minutes per liter. The highest raw water flow and lowest clearwell depth was at 1,800 hours. Therefore, C × T = (44 ft × 30 ft × 8.3 ft [water depth, WD])(7.48)(0.34 baffling factor)(1.6 mg/L)/300 gpm = 148.6 mg-min/L.

Calculate the CT for a clear well contact basin with a baffling factor of 0.5, a capacity of 2.4 million gallons, the plant flow of 5 mgd, and a chlorine residual at the basin outlet of 0.9 mg/L. Detention time = Volume/Flow, so T = 2.4 mg/5 mgd = 0.48 d x 1440 min/d = 691.2 min. Now multiply by 0.5 baffling factor so the allowed detention time is 345.6 min. The CT is calculated by multiplying this by the residual. CT = 0.9 mg/L x 345.6min = 311 min-mg/L.

Lime Softening

Total hardness (mg/L $CaCO_3$) = carbonate hardness (mg/L $CaCO_3$) + noncarbonate hardness (mg/L $CaCO_3$), carbonate hardness = total alkalinity What is the noncarbonated hardness if the alkalinity is 120 mg/L and total hardness is 147 mg/L? Total hardness – total alkalinity = noncarbonated hardness. 147 – 120 = 27 mg/L noncarbonated hardness

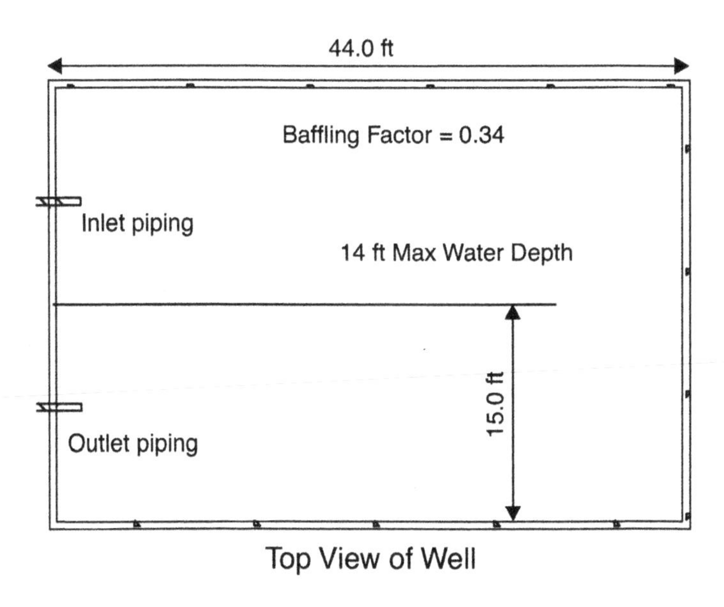

Figure B-6 Typical clearwell arrangement. Note baffle wall, which effectively increases the travel distance of the water and therefore the disinfectant contact time.

Table B-1 Plant operating data for $C \times T$ calculation

Time	Clearwell Depth	Cl_2 Residual	Raw Water Flow, *gpm*	Time	Clearwell Depth	Cl_2 Residual	Raw Water Flow, *gpm*
0000	11.0	1.5	220	1200	9.6	1.5	175
0100	11.0	1.5	220	1300	9.0	1.5	175
0200	10.9	1.5	220	1400	8.7	1.5	175
0300	10.9	1.5	220	1500	8.5	1.5	220
0400	10.6	1.4	220	1600	8.4	1.6	220
0500	10.5	1.5	220	1700	8.3	1.6	220
0600	10.0	1.5	280	1800	8.3	1.6	300
0700	9.8	1.7	270	1900	8.6	1.6	280
0800	9.7	1.7	270	2000	9.0	1.5	280
0900	9.7	1.7	270	2100	9.2	1.5	220
1000	9.9	1.6	220	2200	9.5	1.5	220
1100	10.0	1.6	230	2300	9.9	1.5	220

Calculating the chemical dosages for the excess lime softening process can be involved. These calculations are too lengthy to present in this handbook. Consult the math book, Math for Water Treatment Operators: Practice Problems to Prepare for Water Treatment Operator Certification Exams, in the bibliography section for detailed step by step guidance for these calculations.

Floor loading

The floor loading capacity in the chemical storage room is 438 lb/ft^2. How high can you stack 100-lb bags of soda ash if each bag is 30 in. × 18 in. × 6 in.? Each bag is 2.5 ft × 1.5 ft × 0.5 ft. Surface area of 1 bag is 2.5 ft × 1.5 ft = 3.75 ft^2. Therefore, 3.75 ft^2 × 438 lb/ft^2 = 1,642.5 lb, and 1,642.5 lb × 1 bag/100 lb = 16.4 bags. 16.4 bags × 0.5 ft/bag = 8 feet maximum height.

BIBLIOGRAPHY

Giorgi, John. 2007. *Math for Water Treatment Operators: Practice Problems to Prepare for Water Treatment Operator Certification Exams*, Denver, Colo.: American Water Works Association.

Giorgi, John. 2010. *Water Operator Certification Study Guide*, 6th ed. Denver, Colo.: American Water Works Association.

Appendix C

Water Chemistry

All matter exists in one or more of three states: solid, liquid, or gas. Some substances, such as water, can exist in all three. Simply adding or taking away heat can often bring about the transition from one state to another. Water is a prime example. If we add heat to ice, it will eventually change to a liquid state. This is a physical property of water. We can reverse this change simply by taking away enough heat to freeze the water.

Not all substances will behave similarly. Add heat to a solid piece of wood and it will not normally liquefy—it will char and then burn. This describes a chemical reaction. Once this change has taken place, we cannot go back to the piece of wood. The change was irreversible. The composition and properties of the wood have changed. The study of these types of changes and the nature of compositions of materials is called chemistry.

All of the solid, liquid, and gaseous matter that we know of is composed of one or more elements. There are more than 100 known elements; all of them are listed in the periodic table of the elements (Figure C-1). Table C-1 lists some of the more common elements and their properties. For most purposes, it is best to think of elements as the basic form of matter—those that cannot be broken down further into smaller components by ordinary means. When these elements combine, they form compounds. For example, water is a compound formed from the elements of hydrogen and oxygen. Compounds can usually be broken down to their individual elements.

NOTATION AND FORMULA FORMATS

Every element is represented by a symbol, which is a kind of shorthand to make it easier to read and display chemical formulas. Chemical formulas tell us the number of atoms of each element that make up a compound. For example, the symbol for hydrogen is H and the symbol for oxygen is O. When put together in the ratio of two atoms of hydrogen to one atom of oxygen (H_2O), they represent the formula for water. The formula for carbon dioxide is CO_2 and shows that the compound is made up of one carbon atom and two oxygen atoms.

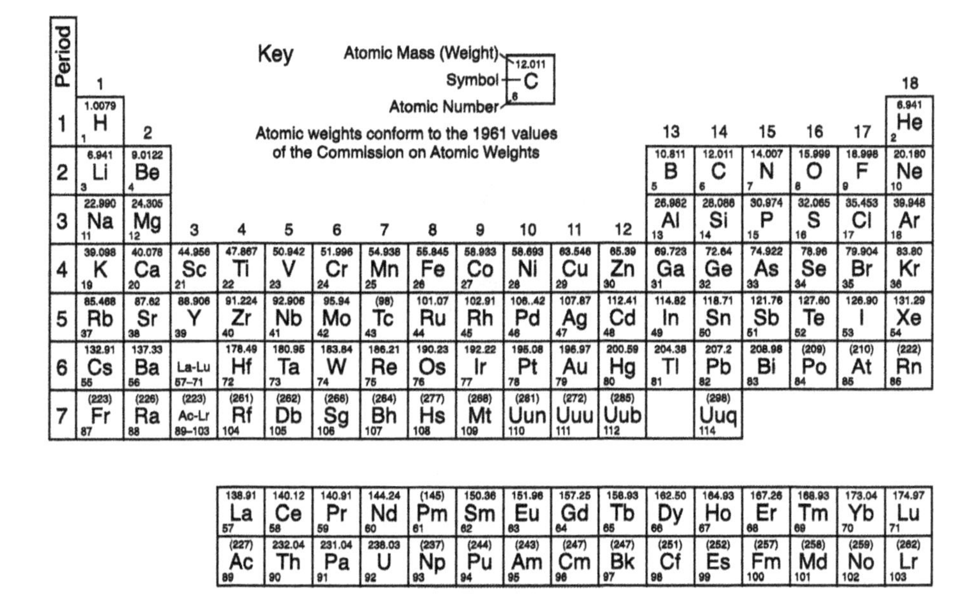

Figure C-1 Periodic table of elements

Table C-1 Chemical names and symbols for some elements common to water treatment

Element	Chemical Symbol	Atomic Weight	Atomic Number	Valence
Aluminum	Al	26.98	13	3
Calcium	Ca	40.08	20	2
Carbon	C	12.0	6	2,4
Chlorine	Cl	35.45	17	1,3,5,7
Copper	Cu	63.54	29	1,2
Fluorine	F	19	9	1
Hydrogen	H	1.008	1	1
Iron	Fe	55.85	26	2,3
Magnesium	Mg	24.31	12	2
Manganese	Mn	54.99	25	2,3,4,6,7
Nitrogen	N	14.01	7	3,5
Oxygen	O	16	8	2
Phosphorus	P	30.98	15	3,5
Potassium	K	39.1	19	1
Sodium	Na	22.99	11	1
Sulfur	S	32.06	16	2,4,6

ATOMIC WEIGHT AND MOLECULAR WEIGHT

Each element has an atomic weight, or formula weight, which helps to further determine the properties of that element. Referring to Table C-1, we see that the atomic weight of hydrogen is 1 and the atomic weight of oxygen is 16. When compounds form, the formula for the compound can be used to determine the molecular weight of the compound. In the example for water, two hydrogens and one oxygen combine to form a compound that has a molecular weight of 18. The percentage of weight that an element contributes to the overall molecule is determined by dividing the elemental weight by the total molecular weight. The percentage of hydrogen in the water molecule is $2/18$, or about 11 percent.

A molecule is the smallest unit of a compound. It is usually made up of several elements in combination but can be made up of two or more of the same elements. The example above shows a molecule of water being made up of two elements. Chlorine, Cl, is an element but it usually exists as Cl_2, the compound called chlorine, which is a molecule. Using Table C-1, we see that the element Cl has a weight of about 35.5, but the chlorine molecule has an atomic weight of 70.

CHEMICAL LAWS, FORMULAS, CALCULATIONS, AND EQUATIONS

Chemistry has laws—predictable results for situations that never change. These laws can help us to make assumptions whenever we work with chemicals in the plant. One law, the Law of Conservation of Matter, states that matter can neither be created nor destroyed, only changed from one form to another. For example, if you bring 1 oz of water to a boil, it eventually turns to steam. The same amount of water is still available, just in a different form.

Other laws of chemistry, such as the Law of Definite Proportions and the Law of Multiple Proportions, state that "if a given compound is formed by the combination of elements, the proportion, by weight, of the element or elements making up the compound is always the same regardless of the method used to form the compound." Also, if "two or more compounds are composed of the same elements, the different weights of one element which combine with a fixed weight of another stand to each other in the ratio of small whole numbers." For example, water (H_2O) and hydrogen peroxide (H_2O_2) are made up of the same elements. Each compound has two hydrogens, but the number of oxygens in the peroxide is twice that of the number found in water. Therefore, the weight of oxygen found in the peroxide is exactly twice the amount found in water.

The Law of Combining Weight states that in every compound, the proportions, by weight, of each element may be expressed by a definite number (atomic weight) or an integral multiple of that number. This is an extension of the Law of Multiple Proportions.

These laws can be used to construct chemical equations, which are used to perform mathematical calculations that predict reactions. Chemical equations are simple expressions that consist of chemical formulas that indicate chemical changes, or reactions, that take place when we bring compounds together. We can show, for example, what happens when chlorine gas is put into water. The equation is

$$Cl_2 + H_2O \rightarrow HOCl + HCl$$

and tells us that a reaction took place and that reaction produced hypochlorous acid and hydrochloric acid. It also tells us that it takes one molecule of each of the reactants to produce one molecule of each of the products. We do not sometimes get two molecules of HOCl—we always get one. We also see that matter has been conserved: there are two chlorines on the left side of the equation and two on the right. Hydrogen and oxygen have also been conserved, although changed.

Sometimes the examples are not so simple—they must be further refined so that none of the laws of chemistry are violated. For example, we know that iron, if exposed to moisture, will oxidize into rust, which is Fe_2O_3. Iron is Fe, and oxygen, like chlorine, exists as a molecule consisting of two elements and is written as O_2. So, if we combine the two compounds that make rust—iron and oxygen—we should get a formula that works. But when we do, it looks like this:

$$Fe + O_2 \rightarrow Fe_2O_3$$

We know that this must be wrong because there is one iron on the left side of the equation but two on the right. Similarly, there are two oxygens on the left and three on the right. Since we cannot create matter, we have to refine, or balance, the equation in order for it to make sense. Balancing equations requires that we start with the correct formulas for the compounds and that we know all of the compounds involved in the reaction. It may seem like a hit-or-miss experience, but with practice, the simple equations can be balanced by anyone. (More complex equations exist, but they are not covered in this text.) In this example, we might start by putting a 2 in front of the Fe_2O_3. Now our formula looks this way:

$$Fe + O_2 \rightarrow 2Fe_2O_3$$

We now have six oxygens on the right side of the equation, but we need six on the left. If we put a 3 in front of the O_2, we would have six there, and the equation would read:

$$Fe + 3O_2 \rightarrow 2Fe_2O_3$$

We are now balanced for oxygen. However, there's still more work to do. The Fe portion of the equation is out of balance—there are four on the right and only one on the left. If we put a 4 in front of the Fe, the equation becomes:

$$4Fe + 3O_2 \rightarrow 2Fe_2O_3$$

The equation is now balanced; it took four molecules of iron to combine with three molecules of oxygen to form two molecules of rust.

This information can be put to practical use predicting the amounts of chemicals needed to produce a desired amount of product. In the example of rusting of iron by oxygen, we saw the proportions of compounds needed to make the product called *rust*. Using Table C-1, we can determine that the atomic weight of iron is 55.85 and the atomic weight of oxygen is 16. Now we can give an account of the weight units of all of the atoms present as follows:

$$\text{four atoms of iron} \rightarrow 4 \times 55.85 = 223.4$$

The O_2 atom is $2 \times 16 = 32$ and there are three of them, so $3 \times 32 = 96$. The iron oxide, or rust, is equivalent to 2×55.85 (two irons) plus 3×16 (three oxygens), or 159.7. Since there are two molecules of rust produced, the overall equivalent for it is 2×159.7, or 319.4. We have learned that 223.4 weight units of iron will combine with 96 weight units of oxygen to form 319.4 weight units of rust. The weight units can be anything: grams, pounds, tons, or ounces, as long as they are the same for each molecule. So, we can say that it takes 223.4 tons of iron and 96 tons of oxygen to form 319.4 tons of rust, and no other proportions will do—it always takes these amounts.

Appendix D

Metric Conversions

Linear Measurement

inch (in.)	×	25.4	=	millimeters (mm)
inch (in.)	×	2.54	=	centimeters (cm)
foot (ft)	×	304.8	=	millimeters (mm)
foot (ft)	×	30.48	=	centimeters (cm)
foot (ft)	×	0.3048	=	meters (m)
yard (yd)	×	0.9144	=	meters (m)
mile (mi)	×	1,609.3	=	meters (m)
mile (mi)	×	1.6093	=	kilometers (km)
millimeter (mm)	×	0.03937	=	inches (in.)
centimeter (cm)	×	0.3937	=	inches (in.)
meter (m)	×	39.3701	=	inches (in.)
meter (m)	×	3.2808	=	feet (ft)
meter (m)	×	1.0936	=	yards (yd)
kilometer (km)	×	0.6214	=	miles (mi)

Area Measurement

square meter (m^2)	×	10,000	=	square centimeters (cm^2)
hectare (ha)	×	10,000	=	square meters (m^2)
square inch ($in.^2$)	×	6.4516	=	square centimeters (cm^2)
square foot (ft^2)	×	0.092903	=	square meters (m^2)
square yard (yd^2)	×	0.8361	=	square meters (m^2)
acre	×	0.004047	=	square kilometers (km^2)
acre	×	0.4047	=	hectares (ha)
square mile (mi^2)	×	2.59	=	square kilometers (km^2)
square centimeter (cm^2)	×	0.16	=	square inches ($in.^2$)
square meters (m^2)	×	10.7639	=	square feet (ft^2)
square meters (m^2)	×	1.1960	=	square yards (yd^2)
hectare (ha)	×	2.471	=	acres
square kilometer (km^2)	×	247.1054	=	acres
square kilometer (km^2)	×	0.3861	=	square miles (mi^2)

Volume Measurement

cubic inch (in.^{3})	×	16.3871	=	cubic centimeters (cm^3)
cubic foot (ft^3)	×	28,317	=	cubic centimeters cm^3)
cubic foot (ft^3)	×	0.028317	=	cubic meters (m^3)
cubic foot (ft^3)	×	28.317	=	liters (L)
cubic yard (yd^3)	×	0.7646	=	cubic meters (m^3)
acre foot (acre-ft)	×	1233.48	=	cubic meters (m^3)
ounce (US fluid) (oz)	×	0.029573	=	liters (L)
quart (liquid) (qt)	×	946.9	=	milliliters (mL)
quart (liquid) (qt)	×	0.9463	=	liters (L)
gallon (gal)	×	3.7854	=	liters (L)
gallon (gal)	×	0.0037854	=	cubic meters (m^3)
peck (pk)	×	0.881	=	decaliters (dL)
bushel (bu)	×	0.3524	=	hectoliters (hL)
cubic centimeters (cm^3)	×	0.061	=	cubic inches (in.^{3})
cubic meter (m^3)	×	35.3183	=	cubic feet (ft^3)
cubic meter (m^3)	×	1.3079	=	cubic yards (yd^3)
cubic meter (m^3)	×	264.2	=	gallons (gal)
cubic meter (m^3)	×	0.000811	=	acre-feet (acre-ft)
liter (L)	×	1.0567	=	quart (liquid) (qt)
liter (L)	×	0.264	=	gallons (gal)
liter (L)	×	0.0353	=	cubic feet (ft^3)
decaliter (dL)	×	2.6417	=	gallons (gal)
decaliter (dL)	×	1.135	=	pecks (pk)
hectoliter (hL)	×	3.531	=	cubic feet (ft^3)
hectoliter (hL)	×	2.84	=	bushels (bu)
hectoliter (hL)	×	0.131	=	cubic yards (yd^3)
hectoliter (hL)	×	26.42	=	gallons (gal)

(Note: US gallons are listed above.)

Pressure Measurement

pounds/square inch (psi)	×	6.8948	=	kilopascals (kPa)
pounds/square inch (psi)	×	0.00689	=	pascals (Pa)
pounds/square inch (psi)	×	0.070307	=	kilograms/square centimeter (kg/cm^2)
pounds/square foot (lb/ft^2)	×	47.8803	=	pascals (Pa)
pounds/square foot (lb/ft^2)	×	0.000488	=	kilograms/square centimeter (kg/cm^2)
pounds/square foot (lb/ft^2)	×	4.8824	=	kilograms/square meter (kg/m^2)
inches of mercury	×	3,376.8	=	pascals (Pa)
inches of water	×	248.84	=	pascals (Pa)
bar	×	100,000	=	newtons per square meter (N/m^2)
pascals (Pa)	×	1	=	newtons per square meter (N/m^2)
pascal (Pa)	×	0.000145	=	pounds/square inch (psi)
kilopascals (kPa)	×	0.145	=	pounds/square inch (psi)
pascal (Pa)	×	0.000296	=	inches of mercury (at 60°F)

kilogram/square centimeter (k/cm^2)	×	14.22	=	pounds/square inch (psi)
kilogram/square centimeter (k/cm^2)	×	28.959	=	inches of mercury (at 60°F)
kilogram/square meter (k/m^2)	×	0.2048	=	pounds per square foot (lb/ft^2)
centimeters of mercury	×	0.4461	=	feet of water

Weight Measurement

ounce (oz)	×	28.3495	=	grams (g)
pound (lb)	×	0.045359	=	grams (g)
pound (lb)	×	0.4536	=	kilograms (kg)
ton (short)	×	0.9072	=	megagrams (metric ton)
pounds/cubic foot (lb/ft^3)	×	16.02	=	grams per liter (g/L)
pounds/million gallons (lb/mil gal)	×	0.1198	=	grams per cubic meter (g/m^3)
gram (g)	×	15.4324	=	grains (gr)
gram (g)	×	0.0353	=	ounces (oz)
gram (g)	×	0.0022	=	pounds (lb)
kilograms (kg)	×	2.2046	=	pounds (lb)
kilograms (kg)	×	0.0011	=	tons (short)
megagram (metric ton)	×	1.1023	=	tons (short)
grams/liter (g/L)	×	0.0624	=	pounds per cubic foot (lb/ft^3)
grams/cubic meter (g/m^3)	×	8.3454	=	pounds/million gallons (lb/mil gal)

Flow Rates

gallons/second (gps)	×	3.785	=	liters per second (L/sec)
gallons/minute (gpm)	×	0.00006308	=	cubic meters per second (m^3/sec)
gallons/minute (gpm)	×	0.06308	=	liters per second (L/sec)
gallons/hour (gph)	×	0.003785	=	cubic meters per hour (m^3/h)
gallons/day (gpd)	×	0.000003785	=	million liters per day (ML/d)
gallons/day (gpd)	×	0.003785	=	cubic meters per day (m^3/d)
cubic feet/second (ft^3/sec)	×	0.028317	=	cubic meters per second (m^3/sec)
cubic feet/second (ft^3/sec)	×	1,699	=	liters per minute (L/min)
cubic feet/minute (ft^3/min)	×	472	=	cubic centimeters per second (cm^3/sec)
cubic feet/minute (ft^3/min)	×	0.472	=	liters per second (L/sec)
cubic feet/minute (ft^3/min)	×	1.6990	=	cubic meters per hour (m^3/h)
million gallons/day (mgd)	×	43.8126	=	liters per second (L/sec)
million gallons/day (mgd)	×	0.003785	=	cubic meters per day (m^3/d)
million gallons/day (mgd)	×	0.043813	=	cubic meters per second (m^3/sec)
gallons/square foot (gal/ft^2)	×	40.74	=	liters per square meter (L/m^2)
gallons/acre/day (gal/acre/d)	×	0.0094	=	cubic meters/hectare/day ($m^3/ha/d$)
gallons/square foot/day ($gal/ft^2/d$)	×	0.0407	=	cubic meters/square meter/day ($m^3/m^2/d$)
gallons/square foot/day ($gal/ft^2/d$)	×	0.0283	=	liters/square meter/day ($L/m^2/d$)
gallons/square foot/minute ($gal/ft^2/min$)	×	2.444	=	cubic meters/square meter/hour ($m^3/m^2/h$) = m/h
gallons/square foot/minute ($gal/ft^2/min$)	×	0.679	=	liters/square meter/second ($L/m^2/sec$)
gallons/square foot/minute ($gal/ft^2/min$)	×	40.7458	=	liters/square meter/minute ($L/m^2/min$)
gallons/capita/day (gpcd)	×	3.785	=	liters/day/capita (L/d per capita)

liters/second (L/sec)	×	22,824.5	=	gallons per day (gpd)
liters/second (L/sec)	×	0.0228	=	million gallons per day (mgd)
liters/second (L/sec)	×	15.8508	=	gallons per minute (gpm)
liters/second (L/sec)	×	2.119	=	cubic feet per minute (ft^3/min)
liters/minute (L/min)	×	0.0005886	=	cubic feet per second (ft^3/sec)
cubic centimeters/second (cm^3/sec)	×	0.0021	=	cubic feet per minute (ft^3/min)
cubic meters/second (m^3/sec)	×	35.3147	=	cubic feet per second (ft^3/sec)
cubic meters/second (m^3/sec)	×	22.8245	=	million gallons per day (mgd)
cubic meters/second (m^3/sec)	×	15,850.3	=	gallons per minute (gpm)
cubic meters/hour (m^3/sec)	×	0.5886	=	cubic feet per minute (ft^3/min)
cubic meters/hour (m^3/sec)	×	4.403	=	gallons per minute (gpm)
cubic meters/day (m^3/d)	×	264.1720	=	gallons per day (gpd)
cubic meters/day (m^3/d)	×	0.00026417	=	million gallons per day (mgd)
cubic meters/hectare/day (m^3/ha/d)	×	106.9064	=	gallons per acre per day (gal/acre/d)
cubic meters/sq meter/day (m^3/m^2/d)	×	24.5424	=	gallons/square foot/day (gal/ft^2/d)
liters/square meter/minute (L/m^2/min)	×	0.0245	=	gallons/square foot/minute (gal/ft^2/min)
liters/square meter/minute (L/m^2/min)	×	35.3420	=	gallons/square foot/day (gal/ft^2/d)

Work, Heat, and Energy

British thermal units (Btu)	× 1.0551	=	kilojoules (kJ)
British thermal units (Btu)	× 0.2520	=	kilogram-calories (kg-cal)
foot-pound (force) (ft-lb)	× 1.3558	=	joules (J)
horsepower-hour (hp-h)	× 2.6845	=	megajoules (MJ)
watt-second (W-sec)	× 1.000	=	joules (J)
watt-hour (W-h)	× 3.600	=	kilojoules (kJ)
kilowatt-hour (kW-h)	× 3,600	=	kilojoules (kJ)
kilowatt-hour (kW-h)	× 3,600,000	=	joules (J)
British thermal units per pound (Btu/lb)	× 0.5555	=	kilogram-calories per kilogram (kg-cal/kg)
British thermal units per cubic foot (Btu/ft^3)	× 8.8987	=	kilogram-calories/cubic meter (kg-cal/m^3)
kilojoule (kJ)	× 0.9478	=	British thermal units (Btu)
kilojoule (kJ)	× 0.00027778	=	kilowatt-hours (kW-h)
kilojoule (kJ)	× 0.2778	=	watt-hours (W-h)
joule (J)	× 0.7376	=	foot-pounds (ft-lb)
joule (J)	× 1.0000	=	watt-seconds (W-sec)
joule (J)	× 0.2399	=	calories (cal)
megajoule (MJ)	× 0.3725	=	horsepower-hour (hp-h)
kilogram-calories (kg-cal)	× 3.9685	=	British thermal units (Btu)
kilogram-calories per kilogram (kg-cal/kg)	× 1.8000	=	British thermal units per pound (Btu/lb)
kilogram-calories per liter (kg-cal/L)	× 112.37	=	British thermal units per cubic foot (Btu/ft^3)
kilogram-calories/cubic meter (kg-cal/m^3)	× 0.1124	=	British thermal units per cubic foot (Btu/ft^3)

Velocity, Acceleration, and Force

feet per minute (ft/min)	×	18.2880	=	meters per hour (m/h)
feet per hour (ft/h)	×	0.3048	=	meters per hour (m/h)
miles per hour (mph)	×	44.7	=	centimeters per second (cm/sec)
miles per hour (mph)	×	26.82	=	meters per minute (m/min)
miles per hour (mph)	×	1.609	=	kilometers per hour (km/h)
feet/second/second (ft/sec^2)	×	0.3048	=	meters/second/second (m/sec^2)
inches/second/second (in./sec^2)	×	0.0254	=	meters/second/second (m/sec^2)
pound force (lbf)	×	4.44482	=	newtons (N)
centimeters/second (cm/sec)	×	0.0224	=	miles per hour (mph)
meters/second (m/sec)	×	3.2808	=	feet per second (ft/sec)
meters/minute (m/min)	×	0.0373	=	miles per hour (mph)
meters per hour (m/h)	×	0.0547	=	feet per minute (ft/min)
meters per hour (m/h)	×	3.2808	=	feet per hour (ft/h)
kilometers/second (km/sec)	×	2.2369	=	miles per hour (mph)
kilometers/hour (km/h)	×	0.0103	=	miles per hour (mph)
meters/second/second (m/sec^2)	×	3.2808	=	feet/second/second (ft/sec^2)
meters/second/second (m/sec^2)	×	39.3701	=	inches/second/second (in./sec^2)
newtons (N)	×	0.2248	=	pounds force (lbf)

Glossary

activated alumina The chemical compound aluminum oxide, which is used to remove fluoride and arsenic from water by adsorption.

activated carbon A highly adsorptive material used to remove organic substances from water. (See adsorption.)

activated silica A coagulant aid used to form a denser, stronger floc.

adsorbent Any material, such as activated carbon, used to adsorb substances from water.

adsorption The water treatment process used primarily to remove organic contaminants from water. Adsorption involves the adhesion of the contaminants to an adsorbent such as activated carbon.

adsorptive media Specially designed and engineered filter material that has unique properties to attract specific ions from water. Operating conditions may need to be adjusted to optimize the performance of these systems.

aeration The process of bringing water and air into close contact to remove or modify constituents in the water.

agglomeration The action of microfloc particles colliding and sticking together to form larger settleable floc particles.

air binding A condition that occurs in filters when air comes out of solution as a result of pressure decreases and temperature increases. The air clogs the voids between the media grains, which causes increased head loss through the filter and shorter filter runs.

air scouring The practice of admitting air through the underdrain system to ensure complete cleaning of media during filter backwash. Normally an alternative to using a surface wash system.

air stripper A packed-tower aerator consisting of a cylindrical tank filled with a packing material made of plastic or other material. Water is usually distributed over the packing at the top of the tank and air is forced in at the bottom using a blower.

alum The most common chemical used for coagulation. It is also called aluminum sulfate.

aluminum sulfate See *alum*.

anionic Having a negative ionic charge.

anode Positive end (pole) of an electrolytic system.

arching A condition that occurs when dry chemicals bridge the opening from the hopper to the dry feeder, clogging the hopper.

auxiliary scour See *filter agitation*.

backwash The reversal of flow through a filter to remove the material trapped on and between the grains of filter media.

baffle A metal, wooden, or plastic plate installed in a flow of water to slow the water velocity and provide a uniform distribution of flow.

bar screen A series of straight steel bars welded at their ends to horizontal steel beams, forming a grid. Bar screens are placed on intakes or in waterways to remove large debris.

bed life The time it takes for a bed of adsorbent to lose its adsorptive capacity. When this occurs, the bed must be replaced with fresh adsorbent.

bivalent ion An ion that has a valence charge of two. The charge can be positive or negative.

body feed In diatomaceous earth filters, the continuous addition of diatomaceous earth during the filtering cycle to provide a fresh filtering surface as the suspended material clogs the precoat.

breakpoint The point at which the chlorine dosage has satisfied the chlorine demand.

breakthrough The point in a filtering cycle at which turbidity-causing material starts to pass through the filter.

calcium carbonate ($CaCO_3$) The principal hardness- and scale-causing compound in water.

calcium hardness The portion of total hardness caused by calcium compounds such as calcium carbonate and calcium sulfate.

carbon dioxide (CO_2) A common gas in the atmosphere that is very soluble in water. High concentrations in water can cause the water to be corrosive. Carbon dioxide is added to water after the lime-softening process to lower the pH in order to reduce calcium carbonate scale formation. This process is known as recarbonation.

carbonate hardness Hardness caused primarily by compounds containing carbonate (CO_3), such as calcium carbonate and magnesium carbonate.

carcinogen A chemical compound that can cause cancer in animals or humans.

cathode The negative end (pole) of an electrolytic system.

cation exchange Ion exchange involving ions that have positive charges, such as calcium and sodium.

cationic polyelectrolyte Polyelectrolyte that forms positively charged ions when dissolved in water.

chlorination The process of adding chlorine to water to kill disease-causing organisms or to act as an oxidizing agent.

chlorinator Any device that is used to add chlorine to water.

clarification Any process or combination of processes that reduces the amount of suspended matter in water.

clarifier See *sedimentation basin.*

coagulant A chemical used in water treatment for coagulation. Common examples are aluminum sulfate and ferric sulfate.

coagulant aid A chemical added during coagulation to improve the process by stimulating floc formation or by strengthening the floc so it holds together better.

coagulation The water treatment process that causes very small suspended particles to attract one another and form larger particles. This is accomplished by the addition of a chemical, called a coagulant, that neutralizes the electrostatic charges on the particles that cause them to repel each other.

coagulation–flocculation The water treatment process that converts small particles of suspended solids into larger, more settleable clumps.

colloidal solid Finely divided solid that will not settle out of water for very long periods of time unless the coagulation–flocculation process is used.

colorimetric A chemical analysis of substances in water using intensity of color (with a specific dye) to determine the concentration.

combined chlorine residual The chlorine residual produced by the reaction of chlorine with substances in the water. Because the chlorine is "combined," it is not as effective a disinfectant as free chlorine residual.

contactor A vertical, steel cylindrical pressure vessel used to hold an activated carbon bed.

conventional filtration A term that describes the treatment process used by most US surface water systems, consisting of the steps of coagulation, flocculation, sedimentation, and filtration.

corrosion The gradual deterioration or destruction of a substance or material by chemical action. The action proceeds inward from the surface.

corrosive Tending to deteriorate material, such as pipe, through electrochemical processes.

coupon test A method of determining the rate of corrosion or scale formation by placing metal strips (coupons) of a known weight in the pipe and examining them for corrosion after a period of time.

***C* × *T* value** The product of the residual disinfectant concentration *C*, in milligrams per litre, and the corresponding disinfectant contact time *T*, in minutes. Minimum *C* × *T* values are specified by the Surface Water Treatment Rule as a means of ensuring adequate kill or inactivation of pathogenic microorganisms in water.

density current A flow of water that moves through a larger body of water, such as a reservoir or sedimentation basin, and does not become mixed with the other water because of a density difference. This difference usually occurs because the incoming water has a different temperature or suspended solids content than the water body.

destratification Use of a method to prevent a lake or reservoir from becoming stratified. Typically consists of releasing diffused compressed air at a low point on the lake bottom.

detention time The average length of time a drop of water or a suspended particle remains in a tank or chamber. Mathematically, it is the volume of water in the tank divided by the flow rate through the tank.

dewatering (of reservoirs) A physical method for controlling aquatic plants in which a water body is completely or partially drained and the plants allowed to die.

dewatering (of sludge) A process to remove a portion of water from sludge.

diaphragm-type metering pump A pump in which a flexible rubber, plastic, or metal diaphragm is fastened at the edges in a vertical cylinder. As the diaphragm is pulled back, suction is exerted and the liquid is drawn into the pump. When it is pushed forward, the liquid is discharged.

diatomaceous earth filter A pressure filter using a medium made from diatoms. The water is forced through the diatomaceous earth by pumping.

diffuser (1) Section of a perforated pipe or porous plates used to inject a gas, such as carbon dioxide or air, under pressure into water. (2) A type of pump.

direct filtration A filtration method that includes coagulation, flocculation, and filtration but excludes sedimentation. Only applicable to raw water relatively low in turbidity because all suspended matter must be trapped by the filters.

disinfectant residual An excess of chlorine left in water after treatment. The presence of residuals indicates that an adequate amount of chlorine has been added at the treatment stage to ensure completion of all reactions with some chlorine remaining.

disinfection The water treatment process that kills disease-causing organisms in water, usually by the addition of chlorine.

disinfection by-products (DBPs) New chemical compounds that are formed by the reaction of disinfectants with organic compounds in water. At high concentrations, many disinfection by-products are considered a danger to human health.

dissolved air flotation A clarification process in which gas bubbles are generated in a basin so that they will attach to solid particles to cause them to rise to the surface. The sludge that accumulates on the surface is then periodically removed by flooding or mechanical scraping.

dissolved solid Any material that is dissolved in water and can be recovered by evaporating the water after filtering the suspended material.

dual-media filtration A filtration method designed to operate at a higher rate by using two different types of filter media, usually sand and finely granulated anthracite.

eductor A device used to mix a chemical with water. The water is forced through a constricted section of pipe (venturi) to create a low pressure, which allows the chemical to be drawn into the stream of water.

effluent launder A trough that collects the water flowing from a basin (effluent) and transports it to the effluent piping system.

empty bed contact time (EBCT) The volume of the tank holding an *activated carbon* bed, divided by the flow rate of water. The EBCT is expressed in minutes and corresponds to the detention time in a sedimentation basin.

epilimnion The upper, warmer layer of water in a stratified lake.

evaporator A device used to increase release of chlorine gas from a container by heating the liquid chlorine.

excess-lime treatment A modification of the lime–soda ash method that uses additional lime to remove magnesium compounds.

filter agitation A method used to achieve more effective cleaning of a filter bed. The system typically uses nozzles attached to a fixed or rotating pipe installed just above the filter media. Water or an air–water mixture is fed through the nozzles at high pressure to help agitate the media and break loose accumulated suspended matter. It can also be called auxiliary scour or surface washing.

filter sand Sand that is prepared according to detailed specifications for use in filters.

filter tank The concrete or steel basin that contains the filter media, gravel support bed, underdrain, and wash-water troughs.

filtration The water treatment process involving the removal of suspended matter by passing the water through a porous medium such as sand.

five haloacetic acids (HAA5) The sum of five specific haloacetic acids, which are mono-, di-, and tri-chloroacetic acids plus mono- and dibromoacetic acids.

flash mixing See *rapid mixing.*

floc Collections of smaller particles (such as silt, organic matter, and micro-organisms) that have come together (agglomerated) into larger, more settleable particles as a result of the coagulation–flocculation process.

flocculation The water treatment process, following coagulation, that uses gentle stirring to bring suspended particles together so that they will form larger, more settleable clumps called floc.

flow measurement A measurement of the volume of water flowing through a given point in a given amount of time.

flow proportional control A method of controlling chemical feed rates by having the feed rate increase or decrease as the flow increases or decreases.

flow tube One type of primary element used in a pressure-differential meter. It measures flow velocity based on the amount of pressure drop through the tube. It is similar to a venturi tube.

fluoridation The water treatment process in which a chemical is added to the water to increase the concentration of fluoride ions to an optimal level. The purpose of fluoridation is to reduce the incidence of dental cavities in children.

fluorosis Staining or pitting of the teeth due to excessive amounts of fluoride in the water.

fluosilicic acid A strongly acidic liquid used to fluoridate drinking water.

free chlorine residual The residual formed once all the chlorine demand has been satisfied. The chlorine is not combined with other constituents in the water and is free to kill microorganisms.

galvanic corrosion A form of localized corrosion caused by the connection of dissimilar metals in an electrolyte such as water.

galvanic series A listing of metals and alloys according to their corrosion potential.

granular activated carbon (GAC) Activated carbon in a granular form, which is used in a bed, much like a conventional filter, to adsorb organic substances from water.

gravel bed Layers of gravel of specific sizes that support the filter media and help distribute the backwash water uniformly.

gravimetric dry feeder See *gravimetric feeder.*

gravimetric feeder Chemical feeder that adds specific weights of dry chemical.

greensand A naturally occurring material glauconite (also called manganese greensand) that can be classified to produce an excellent filtration media and has a special property when coated with manganese oxide to attract iron and manganese from water.

hardness A characteristic of water, caused primarily by the salts of calcium and magnesium. Hardness causes deposition of scale in boilers, damage in some industrial processes, and sometimes objectionable taste.

head loss The amount of energy used by water in moving from one point to another.

humic substance Material resulting from the decay of leaves and other plant matter.

hydrogen sulfide (H_2S) A toxic gas produced by the anaerobic decomposition of organic matter and by sulfate-reducing bacteria. Hydrogen sulfide has a very noticeable rotten-egg odor.

hypochlorination Chlorination using solutions of calcium hypochlorite or sodium hypochlorite.

hypolimnion The lower layer of water in a stratified lake. The water temperature is near 39.2°F (4°C), at which water attains its maximum density.

inlet zone The initial zone in a sedimentation basin. It decreases the velocity of the incoming water and distributes it evenly across the basin.

ion exchange process A process used to remove hardness from water that depends on special materials known as resins. The resins trade nonhardness-causing ions (usually sodium) for the hardness-causing ions calcium and magnesium. The process removes practically all the hardness from water.

ion exchange water softener A treatment unit that removes calcium and magnesium from water using ion exchange resins.

iron An abundant element found naturally in the earth. As a result, dissolved iron is found in most water supplies. When the concentration of iron exceeds 0.3 mg/L, it causes red stains on plumbing fixtures and other items in contact with the water. Dissolved iron can also be present in water as a result of corrosion of cast-iron or steel pipes. This is usually the cause of red-water problems.

iron bacteria Bacteria that use dissolved iron as an energy source. They can create serious problems in a water system because they form large, slimy masses that clog well screens, pumps, and other equipment.

jar test A laboratory procedure for evaluating coagulation, flocculation, and sedimentation processes. Used to estimate the proper coagulant dosage.

jar test calibration The establishment of jar testing mixing and settling conditions so that the results of bench-scale jar tests will give results for the full-scale treatment plant.

lamellar plates A series of thin, parallel plates installed at a 45° angle for shallow-depth sedimentation.

Langelier saturation index (LI) A numerical index that indicates whether calcium carbonate will be deposited or dissolved in a distribution system. The index is also used to indicate the corrosivity of water.

Leopold filter bottom A patented filter underdrain system using a series of perforated vitrified clay blocks with channels to carry the water.

lime–soda ash method A process used to remove carbonate and noncarbonate hardness from water.

loading rate The flow rate per unit area at which the water is passed through a filter or ion exchange unit.

magnesium hardness The portion of total hardness caused by magnesium compounds such as magnesium carbonate and magnesium sulfate.

magnetic flowmeter A flow-measuring device in which the movement of water induces an electrical current proportional to the rate of flow.

manganese An abundant element found naturally in the earth. Dissolved manganese is found in many water supplies. At concentrations above 0.05 mg/L, it causes black stains on plumbing fixtures, laundry, and other items in contact with the water.

manual solution feed A method of feeding a chemical solution for small water systems. The chemical is dissolved in a small plastic tank, transferred to another tank, and fed to the water system by a positive-displacement pump.

maximum contaminant level (MCL) The maximum allowable concentration of a contaminant in drinking water, as established by state and/or federal regulations. Primary MCLs are health related and mandatory. Secondary MCLs are related to the aesthetics of the water and are highly recommended but not required.

membrane processes Water treatment processes in which relatively pure water passes through a porous membrane while particles, molecules, or ions of unwanted matter are excluded. The membrane process used primarily for potable water treatment is reverse osmosis.

metering pump A chemical solution feed pump that adds a measured volume of solution with each stroke or rotation of the pump.

microfloc The initial floc formed immediately after coagulation, composed of small clumps of solids.

microstrainer A rotating drum lined with a finely woven material such as stainless steel. Microstrainers are used to remove algae and small debris before they enter the treatment plant.

milk of lime The lime slurry formed when water is mixed with calcium hydroxide.

mottling The staining of teeth due to excessive amounts of fluoride in the water.

mudball An accumulation of media grains and suspended material that creates clogging problems in filters.

multimedia filter A filtration method designed to operate at a high rate by utilizing three or more different types of filter media. The media types typically used are silica sand, anthracite, and garnet sand.

negative head A condition that can develop in a filter bed when the head loss gets too high. When this occurs, the pressure in the bed can drop to less than atmospheric.

nephelometric turbidimeter An instrument that measures turbidity by measuring the amount of light scattered by turbidity in a water sample. It is the only instrument approved by the US Environmental Protection Agency to measure turbidity in treated drinking water.

nephelometric turbidity unit (ntu) The amount of turbidity in a water sample as measured by a nephelometric turbidimeter.

noncarbonate hardness Hardness caused by the salts of calcium and magnesium.

nonionic polyelectrolyte Polyelectrolyte that forms both positively and negatively charged ions when dissolved in water.

nonsettleable solids Finely divided solids, such as bacteria and fine clay particles, that will stay suspended in water for long periods of time.

ntu See *nephelometric turbidity unit.*

on-line turbidimeter A turbidimeter that continuously samples, monitors, and records turbidity levels in water.

organic substance (organic) A chemical substance of animal or vegetable origin, having carbon in its molecular structure.

orifice plate A type of primary element used in a pressure-differential meter, consisting of a thin plate with a precise hole through the center. Pressure drops as the water passes through the hole.

outlet zone The final zone in a sedimentation basin. It provides a smooth transition from the settling zone to the effluent piping.

overflow weir A steel or fiberglass plate designed to distribute flow evenly. In a sedimentation basin; the weir is attached to the effluent launder.

oxidation (1) The chemical reaction in which the valence of an element increases because of the loss of electrons from that element. (2) The conversion of organic substances to simpler, more stable forms by either chemical or biological means.

oxidize To chemically combine with oxygen.

ozone contactor A tank used to transfer ozone to water. A common type applies ozone under pressure through a porous stone at the bottom of the tank.

ozone generator A device that produces ozone by passing an electrical current through air or oxygen.

packed tower A cylindrical tank containing packing material, with water distributed at the top and airflow introduced from the bottom by a blower. Commonly referred to as an air stripper.

packing material The material placed in a packed tower to provide a very large surface area over which water must pass to attain a high liquid–gas transfer.

pathogen A disease-causing organism.

percolation The movement or flow of water through the pores of soil, usually downward.

permanent hardness Another term for noncarbonate hardness, derived from the fact that the hardness-causing noncarbonate compounds do not precipitate when the water is boiled.

pilot filter A small tube, containing the same media as treatment plant filters, through which flocculated plant water is continuously passed, with a recording turbidimeter continuously monitoring the effluent. The amount of water passing through the pilot filter before turbidity breakthrough can be correlated to the operation of the plant filters under the same coagulant dosage.

pipe lateral system A filter underdrain system using a main pipe (header) with several smaller perforated pipes (laterals) branching from it on both sides.

plain sedimentation The sedimentation of suspended matter without the use of chemicals or other special means.

polyelectrolyte High-molecular-weight, synthetic organic compound that forms ions when dissolved in water. It is also called a polymer.

polymer See *polyelectrolyte.*

porous plate A filter underdrain system using ceramic plates supported above the bottom of the filter tank. This system is often used without a gravel layer so that the plates are directly beneath the filter media.

powdered activated carbon (PAC) Activated carbon in a fine powder form. It is added to water in a slurry form primarily for removing those organic compounds causing tastes and odors.

precipitate (1) A substance separated from a solution or suspension by a chemical reaction. (2) To form such a substance.

precoating The initial step in diatomaceous earth filtration, in which a thin coat of diatomaceous earth is applied to a support surface called a septum. This provides an initial layer of media for the water to pass through.

precursor compound Any of the organic substances that react with chlorine to form trihalomethanes.

preliminary treatment Any physical, chemical, or mechanical process used before the main water treatment processes. It can include screening, presedimentation, and chemical addition. Also called pretreatment.

presedimentation A preliminary treatment process used to remove gravel, sand, and other gritty material from the raw water before it enters the main treatment plant. This is usually done without the use of coagulating chemicals.

presedimentation impoundment A large earthen or concrete basin used for presedimentation of raw water. It is also useful for storage and for reducing the impact of raw-water quality changes on water treatment processes.

pressure-differential meter Any flow-measuring device that creates and measures a difference in pressure proportionate to the rate of flow. Examples include the venturi meter, orifice meter, and flow nozzle.

pressure-sand filter A sand filter placed in a cylindrical steel pressure vessel. The water is forced through the media under pressure.

pretreatment See *preliminary treatment.*

primary drinking water regulations Regulations on drinking water quality that are considered essential for preservation of public health.

quicklime Another name for calcium oxide (CaO), which is used in water softening and stabilization.

radial flow Flow that moves across a basin from the center to the outside edges or vice versa.

radionuclide A radioactive element that is usually regulated in drinking water.

rapid mixing The process of quickly mixing a chemical solution uniformly through the water.

rate-of-flow controller A control valve used to maintain a fairly constant flow through the filter.

reactivate To remove the adsorbed materials from spent activated carbon and restore the carbon's porous structure so that it can be used again. The reactivation process is similar to that used to activate carbon.

recarbonation The reintroduction of carbon dioxide into the water, either during or after lime–soda ash softening, to lower the pH of the water.

red water Rust-colored water resulting from the formation of ferric hydroxide from iron naturally dissolved in the water or from the action of iron bacteria.

regeneration The process of reversing the ion exchange softening reaction of ion exchange materials. Hardness ions are removed from the used materials and replaced with nontroublesome ions, thus rendering the materials fit for reuse in the softening process.

regeneration rate The flow rate per unit area of an ion exchange resin at which the regeneration solution is passed through the resin.

reject water The water that does not pass through a membrane, carries away the rejected matter, and must be disposed of.

residual See *disinfectant residual.*

residual flow control A method of controlling the chlorine feed rate based on the residual chlorine after the chlorine feed point.

resin In water treatment, the synthetic, bead-like material used in the ion exchange process.

reverse osmosis A pressure-drive process in which almost-pure water is passed through a semipermeable membrane. Water is forced through the membrane and most ions (salts) are left behind. The process is principally used for desalination of sea water.

riverbank filtration Extraction of water through the bank of a river or through the bottom using wells or buried collection piping.

rotameter A flow measurement device used for gases.

sand boil The violent washing action in a filter caused by uneven distribution of backwash water.

saturation point The point at which a solution can no longer dissolve any more of a particular chemical. Precipitation of the chemical will occur beyond this point.

saturator A piece of equipment that feeds a sodium fluoride solution into water for fluoridation. A layer of sodium fluoride is placed in a plastic tank and water is allowed to trickle through the layer, forming a solution of constant concentration that is fed to the water system.

schmutzdecke The layer of solids and biological growth that forms on top of a slow sand filter, allowing the filter to remove turbidity effectively without chemical coagulation.

secondary drinking water regulations Regulations developed under the Safe Drinking Water Act that establish maximum levels for substances affecting the aesthetic characteristics (taste, odor, or color) of drinking water.

sedimentation The water treatment process that involves reducing the velocity of water in basins so that the suspended material can settle out by gravity.

sedimentation basin A basin or tank in which water is retained to allow settleable matter, such as floc, to settle by gravity. Also called a settling basin, settling tank, or sedimentation tank.

sedimentation tank See *sedimentation basin.*

sequestering agent A chemical compound such as EDTA (ethylenediaminetetraacetic acid) or certain polymers that chemically tie up (sequester) other compounds or ions so that they cannot be involved in chemical reactions.

settleability test A determination of the settleability of solids in a suspension by measuring the volume of solids settled out of a measured volume of sample in a specified interval of time, usually reported in millilitres per litre.

settling basin See *sedimentation basin.*

settling tank See *sedimentation basin.*

settling zone The zone in a sedimentation basin that provides a calm area so that the suspended matter can settle.

shallow-depth sedimentation A modification of the traditional sedimentation process using inclined tubes or plates to reduce the distance the settling particles must travel to be removed.

short-circuiting A hydraulic condition in a basin in which the actual flow time of water through the basin is less than the design flow time (detention time).

slake The addition of water to quicklime (calcium oxide) to form calcium hydroxide, which can then be used in the softening or stabilization processes.

slaker The part of a quicklime feeder that mixes the quicklime with water to form hydrated lime (calcium hydroxide).

slow sand filtration A filtration process that involves passing raw water through a bed of sand at low velocity, resulting in particulate removal by physical and biological mechanisms.

sludge The accumulated solids separated from water during treatment.

sludge-blanket clarifier See *solids-contact basin.*

sludge blowdown The controlled withdrawal of sludge from a solids-contact basin to maintain the proper level of settled solids in the basin.

sludge zone The bottom zone of a sedimentation basin. It receives and stores the settled particles.

slurry A thin mixture of water and any insoluble material, such as activated carbon.

sodium fluoride A dry chemical used in the fluoridation of drinking water. It is commonly used in saturators.

sodium silicofluoride A dry chemical used in the fluoridation of drinking water. It is derived from fluosilicic acid.

softening The water treatment process that removes calcium and magnesium, the hardness-causing constituents in water.

solids-contact basin A basin in which the coagulation, flocculation, and sedimentation processes are combined. The water flows upward through the basin. It is used primarily in the lime softening of water. It can also be called an upflow clarifier or sludge-blanket clarifier.

solids-contact process A process combining coagulation, flocculation, and sedimentation in one treatment unit in which the flow of water is vertical.

spray tower A tower built around a spray aerator to keep the wind from blowing the spray and to prevent the water from freezing during cold temperatures.

stabilization The water treatment process intended to reduce the corrosive or scale-forming tendencies of water.

static mixer A device designed to produce turbulence and mixing of chemicals with water, by means of fixed sloping vanes within the unit, without the need for any application of power.

sterilization The destruction of all organisms in water.

streaming current monitor An instrument that passes a continuous sample of coagulated water past a streaming current detector (SCD). The measurement is similar in theory to zeta potential determination and provides a reading that can be used to optimize chemical application.

surface overflow rate A measurement of the amount of water leaving a sedimentation tank per square foot of tank surface area.

surface washing See *filter agitation*.

suspended solid Solid organic and inorganic particle that is held in suspension by the action of flowing water.

SUVA Specific UV Absorbance (SUVA) is an analysis test used to estimate the organic content of water. The absorbance is recorded at 254 nm for this measurement.

synthetic organic chemical A carbon-containing chemical that has been manufactured, as opposed to occurring in nature.

synthetic resin See *resin*.

temporary hardness Another term for carbonate hardness, derived from the fact that the hardness-causing carbonate compounds precipitate when water is heated.

terminal head loss The head loss in a filter at which water can no longer be filtered at the desired rate because the suspended matter fills the voids in the filter and greatly increases the resistance to flow (head loss).

thermocline The temperature transition zone in a stratified lake, located between the epilimnion and the hypolimnion.

THM See *trihalomethane*.

titration Process of chemical analysis in which a solution of known concentration is gradually added to the unknown solution from a burette (a long measuring tube with a valve at the bottom) until the equivalence point (end point) is reached. The amount of the unknown substance can then be calculated.

total organic carbon (TOC) The amount of carbon bound in organic compounds in a water sample as determined by a standard laboratory test.

total trihalomethanes (TTHM) The sum of four trihalomethanes, which are chloroform, bromodichloromethane, dibromochloromethane, and bromoform

tracer study A study using a substance that can readily be identified in water (such as a dye) to determine the distribution and rate of flow in a basin, pipe, or channel.

transmitter The part of a pressure-differential meter that measures the signal from the primary element and sends another signal to the receiver.

trihalomethane (THM) A compound formed when natural organic substances from decaying vegetation and soil (such as humic and fulvic acids) react with chlorine.

tube settlers A series of plastic tubes about 2 in. (50 mm) square, used for shallow-depth sedimentation.

tube-settling A shallow-depth sedimentation process that uses a series of inclined tubes.

turbidity A physical characteristic of water making the water appear cloudy. The condition is caused by the presence of suspended matter.

turbulence A flow of water in which there are constant changes in flow velocity and direction resulting in agitation.

ultrasonic flowmeter A water meter that measures flow rate by measuring the difference in the velocity of sound beams directed through the water.

underdrain The bottom part of a filter that collects the filtered water and uniformly distributes the backwash water.

unstable Corrosive or scale forming.

UV disinfection Disinfection using an ultraviolet light.

Van der Waals force The attractive force existing between colloidal particles that allows the coagulation process to take place.

velocity meter A meter that measures water velocity by using a rotor with vanes (such as a propeller). It operates on the principle that the vanes move at about the same velocity as the flowing water.

viscosity The resistance of a fluid to flowing due to internal molecular forces.

viscous Having a sticky quality.

volatile Capable of turning to vapor (evaporating) easily.

volatile organic chemicals (VOCs) A class of manufactured, synthetic chemicals that are generally used as industrial solvents. They are classified as known or suspected carcinogens or as causing other adverse health effects. They are of particular concern to the water supply industry because they have widely been found as contaminants in groundwater sources.

volumetric dry feeder See *volumetric feeder*.

volumetric feeder A chemical feeder that adds specific volumes of dry chemical.

wash-water trough A trough placed above the filter media to collect the backwash water and carry it to the drainage system.

water age The amount of time that drinking water is stored within the distribution network. The total time from entry into the distribution system until it is used.

waterborne disease A disease caused by a waterborne organism or toxic substance.

weighting agent A material, such as bentonite, added to low-turbidity waters to provide additional particles for good floc formation.

weir overflow rate A measurement of the number of gallons per day of water flowing over each foot of weir in a sedimentation tank or circular clarifier. Mathematically, it is the gallons-per-day flow over the weir divided by the total length of the weir in feet.

wheeler bottom A patented filter underdrain system using small porcelain spheres of various sizes in conical depressions.

zeta potential A measurement (in millivolts) of the particle charge strength surrounding colloidal solids. The more negative the number, the stronger the particle charge and the repelling force between particles.

Index

Note: An *f.* following a page number indicates a figure. A *t.* indicates a table.

About the Authors

Nicholas G. Pizzi

Nick Pizzi was vice-president of Environmental Engineering & Technology, Inc., an environmental engineering firm in Twinsburg, Ohio, and Newport News, Virginia. He also owns Aqua Serv Consultants, a business that caters to training and regulatory needs of water plant operators. Previously, he was the assistant commissioner of operations and water quality manager for the Cleveland, Ohio, Division of Water. Mr. Pizzi holds a Class IV water supply license in Ohio and a BA degree in chemistry from St. Vincent College in Latrobe, Pennsylvania. He was chair of the Partnership for Safe Water's Program Effectiveness Assessment Committee and a trustee for the Distribution and Plant Operations Division of American Water Works Association (AWWA). He has served on various Water Research Foundation project advisory committees and as a *Journal AWWA* reviewer. He also served on the *Opflow* editorial advisory board, the AWWA Standards council, and the TEC.

Mr. Pizzi has spent 37 years in the fields of plant and distribution system operations, along with training and consulting to operators of these systems. He has designed and codesigned several training classes for operators in Ohio, published articles in the *Journal – American Water Works Association* and *Opflow*, and made presentations at AWWA annual conferences and AWWA's Water Quality Technology Conference. In 1995, he revised the 12th edition of *Hoover's Water Supply and Treatment* (National Lime Association, Kendall Hunt Publishers, Dubuque, Iowa). In 2004, he published AWWA's "Stories From the Road," a text for water treatment plant operators. He produced several operator guidebooks designed for front line personnel. In all, he has written and/or edited seven AWWA books.

AWWA awarded Mr. Pizzi with the *Opflow* Publication Award in 1997 and with the Ohio Section Education Award. He is a George Warren Fuller Awardee.

William C. Lauer

Mr. Lauer has more than 35 years of experience in drinking water quality and treatment process technology. He is an internationally recognized authority on drinking water health effects and treatment methods and authoring or editing twenty-one books and more than 70 articles on these subjects. He has served on many advisory panels such as: Environmental Protection Agency Blue Ribbon Panel on Best Available Treatment Technologies, National Aeronautics and Space Administration (NASA) Health Effects Issues for International Space Station, the National Academy of Science Health Effects of Recycled Water, and the Government of Singapore Reclaimed Water Expert Panel.

Water treatment process optimization for water quality improvement is a specialty. Mr. Lauer is an expert in water quality issues for water collection, treatment and distribution systems. His projects have included those dealing with drinking water odors, corrosion, cross-connection control, disinfection, discolored water, customer relations, water system management, and particulate removal processes. Mr. Lauer has recently completed serving more than fifteen years as Program Manager for the *Partnership for Safe Water* for AWWA. Mr. Lauer was the staff technical advisor for the development of AWWA utility management standards for; Water Treatment Plant Operation and Management, and Distribution System Operations. These programs seek to improve water and wastewater utility operations and the quality of services.

Mr. Lauer was project manager during design, construction, operation, and testing of Denver's $30 million Direct Potable Reuse Demonstration Project. The 1 mgd advanced treatment plant (referred to as the world's most complex) included: high pH lime clarification, two-stage softening, recarbonation, granular media pressure filtration, selective ion exchange and its associated ammonia recovery and removal process, two-stage granular activated carbon adsorption and a fluidized bed activated carbon regeneration furnace, ultraviolet irradiation disinfection, reverse osmosis, air stripping, ozonation, chlorine dioxide disinfection, chloramine disinfection and micro and ultrafiltration. More than ten years of studies including lifetime whole animal health effects evaluations concluded that the direct conversion of treated secondary wastewater effluent to drinking water quality was feasible. Mr. Lauer is a leading authority in water reuse and as such participates in many national and international water reuse projects.

Mr. Lauer managed the water quality program for the Denver Water Department, serving more than one million customers, and supervised operation of conventional water treatment facilities with a combined treatment capacity of more than 600 million gallons per day. In this regard, he was responsible for the training, education, and management of three water treatment plants. Mr. Lauer managed the water quality laboratory as part of his responsibility. The laboratory employed state-of-the-art testing methods to assess chemical and microbiological safety. Many research projects were conducted under Mr. Lauer's direction to support treatment plant operations, source water protection, and enhance distribution system integrity.